ROADSIDE GEOLOGY OF WASHINGTON

ROADSIDE GEOLOGY
of Washington

David D. Alt
Donald W. Hyndman

MOUNTAIN PRESS PUBLISHING COMPANY
Missoula, Montana

Copyright © 1994
David D. Alt and Donald W. Hydman

Sixteenth Printing, April 2005

Library of Congress Cataloging-in-Publication Data

Alt, David D.
    Roadside geology of Washington

    1. Geology—Washington (State).   I. Hyndman, Donald W.
II. Title.
QE175.A48   1984   55797   84-8409
ISBN 0-87842-160-2

PRINTED IN THE U.S.A.

Mountain Press Publishing Company
P.O. Box 2399 • Missoula, MT 59806
(406) 728-1900

PREFACE

We wrote this book for people who are not geologists, but would like to know something about the rocks and landscapes of Washington. Those who are not geologists often find it difficult to learn much about the subject because the literature of geology is highly technical, and generally available only in major libraries. That is unfortunate because most of that literature reports publicly funded research.

A comprehensive discussion of the detailed geology of Washington would fill a shelf full of books, and this is only one volume. So we focused on the most visible, most significant, and most intriguing aspects of Washington geology, and used a broad brush to paint the picture in bold strokes. We skimmed the cream. Our intention throughout was to enable our readers to recognize the rocks and landscapes of Washington, and understand something of what they mean.

In addition to our own research and our own familiarity with the geology of Washington, we went through the published literature of Washington geology, and tried to distill the essence out of a century of research. We do not provide a bibliography or literature citations in the text because that sort of thing seems inappropriate in a work of this type. So it is impossible for us properly to acknowledge our sources of information except to say that they include practically everyone who has published anything on the geology of Washington. Those people will recognize their own contributions if they read this book, and we want them to know that we recognize them too. We thank them all, and hope they will forgive us for not mentioning each one individually. We also thank Laurie Emmart for the final art work on the maps.

Washington owes much of its existence to the processes of plate tectonics, which were virtually unheard of until the mid-1960's. Much of the literature of Washington geology was published before any geologist knew of plate tectonics, and application of that body of theory to the geology of Washington is still in its early stages. Therefore, large parts of our plate tectonic interpretation of Washington geology are original, and this book is their first publication.

Missoula, Montana
June, 1984

CONTENTS

MAP SYMBOLS

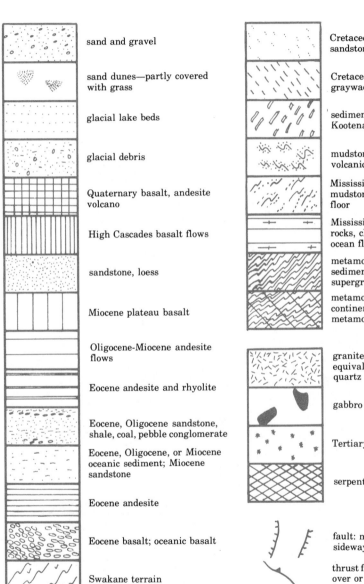

sand and gravel

sand dunes—partly covered with grass

glacial lake beds

glacial debris

Quaternary basalt, andesite volcano

High Cascades basalt flows

sandstone, loess

Miocene plateau basalt

Oligocene-Miocene andesite flows

Eocene andesite and rhyolite

Eocene, Oligocene sandstone, shale, coal, pebble conglomerate

Eocene, Oligocene, or Miocene oceanic sediment; Miocene sandstone

Eocene andesite

Eocene basalt; oceanic basalt

Swakane terrain

Cretaceous to Paleocene sandstone, conglomerate

Cretaceous siltstone, graywacke, mudstone

sedimentary rocks of the Kootenay arc

mudstone, volcanic sandstone, volcanic tuff

Mississippian-Triassic chert, mudstone, limestone: ocean floor

Mississippian-Triassic volcanic rocks, chert, mudstone (old ocean floor)

metamorphosed Precambrian sedimentary rocks of the Belt supergroup

metamorphic rocks of the old continental crust: gneiss, schist, metamorphosed granite

granite and its darker equivalents, granodiorite, quartz diorite

gabbro

Tertiary dunite

serpentinite

fault: movement may be sideways or up and down

thrust fault: points of side thrust over or point in direction of underthrusting

GEOLOGIC TIME SCALE

Rocks	Period Name		Important events in Washington
	Recent		Sea level rose 300 ft. to present level 8000 to 10,000 years ago.
glacial deposits, loess, sand, gravel	Pleistocene		Ice ages; northern third of state covered by continental ice; Puget Sound and Okanogan Valley eroded by ice; Spokane floods scour the Columbia Plateau.
mudflows, andesite, basalt			Mt. Rainier reached maximum size 75,000 years ago; Mt. St. Helens mostly built 2000 years ago to present. Eruption of High Cascades volcanoes.
loess			Winds blow dust east across Columbia Plateau to form Palouse loess.

BEGAN BETWEEN 2 AND 3 MILLION YEARS AGO

gravel, volcanic debris	Pliocene	2-10	Dry climate; Ellensburg formation; alluvial fans.

BEGAN ABOUT 11 MILLION YEARS AGO

flood basalts	Miocene	13-16	Eruption of Columbia River plateau basalts.
clay beds			Wet tropical climate forms Latah formation.
granite			Tatoosh pluton under Mt. Rainier volcano.

BEGAN ABOUT 24 MILLION YEARS AGO

andesite flows, mudflows, sediments, basalt, rhyolite	Oligocene	20-40	Volcanoes active in the Western Cascades; Ohanepecosh, Stevens Ridge, Fifes Peak formations.
sandstone, shale coal			Coastal plains and coastal swamps around North Cascade subcontinent: in Puget Sound lowland and near Cle Elum and Wenatchee.
turbidite sands, ocean-floor basalt	Eocene	40-50	Continental shelf and trench sediments overlying basalt in the Olympic Peninsula. North Cascade micro-continent docks against North America; volcanoes and Republic graben; movement on Straight Creek fault and Newport fault

CENOZOIC

Rocks		Period Name		Important events in Washington

BEGAN ABOUT 60 MILLION YEARS AGO

granite

Intrusion of plutons including Mt. Stuart batholith.

Rocks	Era	Period Name	Age	Important events in Washington
	MESOZOIC	Cretaceous	100	Okanogan micro-continent docks against North America; metamorphism in Okanogan Valley areas.
sandstone, mudstone				Coastal plain sediments of Methow graben deposited; old coastal plain crumpled to become the Kootenay arc.
		Jurassic	150	Ocean-floor rocks of San Juan Islands stuffed into oceanic trench.
		Triassic		

BEGAN ABOUT 240 MILLION YEARS AGO

Rocks	Era	Period Name	Age	Important events in Washington
mudstones, muddy sandstones	PALEOZOIC	Permian Pennsylvanian	300 to	Oceanic sediments in western Washington.
chert, mudstone, limestone, ocean-floor volcanic rocks		Mississippian Devonian Cambrian	600	Sediments deposited in coastal plain in northeastern Washington later deformed to form the Kootenay arc.

BEGAN ABOUT 570 MILLION YEARS AGO

Rocks	Era	Period Name	Age	Important events in Washington
	PRECAMBRIAN		1000 to	Asia (?) splits off from western North America.
metamorphosed mudstones, siltstones		Proterozoic	1400	Deposition of Belt supergroup.
gneiss, schist, granite				Periods of sediment deposition, metamorphism, and igneous intrusion to form the old continental
		Archean	3200	crust of North America.

Major geological regions of Washington

I
INTRODUCTION

A PATCHWORK STATE

In discussing the geology of Washington, we speak mostly of islands that came in from the sea, and how they made our continent grow. Two hundred million years ago, most of Washington was two large islands, each one a scrap of continent, lying somewhere in the vastness of the Pacific Ocean. One after the other, they docked onto the North American continent, each adding its distinctive bit to the complex geologic and geographic mosaic of western North America.

We begin with a brief discussion of the geologic mechanisms that move, assemble, and dismantle continents. Then we briefly outline the sequence of events that assembled the mosaic of pieces that make Washington. More details follow in the chapters dealing with the geologic regions of Washington.

THE EARTH'S OUTER RIND

The Lithosphere

The earth has at its center a metallic core about the size of the moon. Surrounding the core is the mantle, a thick shell of dark green and black rocks called peridotite that comprise the largest part of our planet. The outer rind of the mantle, a shell

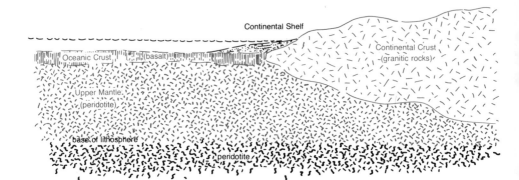

Continental Shelf

Oceanic Crust (basalt)

Continental Crust (granitic rocks)

Upper Mantle (peridotite)

base of lithosphere

peridotite

A cross-section of the earth's lithosphere. Thick continental and thin oceanic crust cover a relatively cold and rigid outer shell of the mantle. The base of the lithosphere lies at a depth of about 60 miles, where the mantle becomes partially molten. The rigid lithosphere moves by slipping on the partially molten rock beneath.

about 60 miles thick, is cooler than the rock below, and therefore more rigid. A thin outer skin called the crust covers the peridotite of the mantle nearly everywhere. Nevertheless, there are a few exposures of mantle rock in Washington, and we will point them out. Geologists use the term lithosphere to refer to the rigid outer shell of the mantle, along with the crust that covers it. Lithosphere covers the entire planet like the rind on a melon.

Continental Crust

Think of a continent as a floating raft about 35 miles thick composed of relatively light rocks, mostly granite and rocks closely related to granite, floating on the much denser peridotite of the mantle. Large areas of continental crustal rocks are exposed along the highways of northern Washington, and we will point them out too. However, in most parts of the continents, as in much of Washington, the granitic crustal rocks lie buried beneath a cover of younger sedimentary and volcanic rocks.

Oceanic Crust

Where no raft of continental granite lies embedded in the

upper part of the lithosphere, the mantle carries a thin skin of oceanic crust. It consists mostly of a black volcanic rock called basalt, which also erupts from volcanoes on the continents. Most people who travel in Washington probably see more basalt than any other kind of rock, and we will point it out in many places.

As you might expect, most of the world's inventory of oceanic crust lies on the seafloor, where two miles of water make direct observation difficult. However, Washington is unusual among land areas in having oceanic instead of continental crust beneath large parts of the state. In much of western Washington, people live and travel in landscapes eroded into oceanic crust, and see roadcuts in rocks that properly belong on the ocean floor. Oceanic crust also exists beneath much of eastern Washington, but it is not exposed at the surface there.

Basalt pillows exposed in the San Juan Islands.

Basalt lava that erupts on the ocean floor behaves very much like molten wax dribbling down the side of a candle. The lava chills against the cold sea water to form a hard skin, and then the molten basalt within bursts through that skin, and extrudes onto the ocean floor as a great cylinder of basalt several feet in diameter. Much of the oceanic crust consists of such lava flows. When we see them exposed in cross-section in roadcuts, the cylinders of basalt look like big greenish black sofa pillows, which explains why geologists call them pillow basalts. People who travel in western Washington see pillow basalts along some of the roads.

PLATE TECTONICS

Although it covers the entire earth, the lithosphere is not continuous like the seamless rind of a melon. Instead, it consists of a mosaic of pieces called plates that fit together like the bones in a skull. About a dozen lithospheric plates of greatly differing sizes completely tile the earth's surface. Plates all move constantly, each one going its own way, each at a different rate. In some places, plates slide past each other along great faults that geologists call transform boundaries. In other places, plates pull directly away from each other at the crests of oceanic ridges. Where two plates collide, one sinks beneath the other to form an oceanic trench that lies parallel to a chain of volcanoes. All those kinds of plate boundaries play a role in the geologic picture of Washington.

Transform Plate Boundaries

The San Andreas fault in California is a good example of a transform boundary between two lithospheric plates that simply slide past each other. In that case, the Pacific plate slips north past the North American plate carrying along a slice of western California. The Queen Charlotte fault off the coast of British Columbia is a similar transform boundary between the Pacific and North American plates. Although neither of those transform plate boundaries enters the state, both fit into the broad picture of Washington geology.

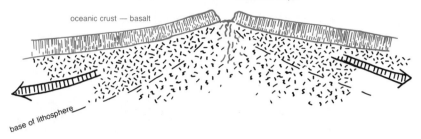

A cross section of an oceanic ridge. As the plates pull away from each other, following the arrows, the mantle beneath partially melts to produce basalt magma, which erupts into the opening gap to form new oceanic crust.

Oceanic Ridges

Plates pull away from each other at the crests of oceanic ridges, broad bulges with a split at the top that wind through the ocean basins of the world in a pattern that crudely suggests the stitching on a baseball. As plates separate at the crest of the oceanic ridge, a fissure opens between them, and molten basalt magma from the mantle wells up through it to form a new basalt lava flow, new oceanic crust.

Some authors have compared what happens at the crest of the oceanic ridge to an imaginary situation in which two ice floes pull away from each other. If water rising into the gap freezes onto the split between the floes, each will grow as fast as they separate. Similarly, oceanic crust forms at the crests of oceanic ridges as fast as the plates separate, typically at a rate of about 2 or 3 inches per year. As the plates with their newly formed oceanic crust move away from each other, they cool off, making the lithosphere beneath them heavier and thicker.

Obviously, if new oceanic crust forms at oceanic ridges, old oceanic crust must be disappearing elsewhere. That happens where plates collide.

Oceanic Trenches and Volcanic Chains

As long as at least one of a pair of colliding plates has oceanic crust on its surface, that plate will sink back into the mantle,

5

leaving the other at the surface. If both colliding plates have oceanic crust on them, the one farthest from the ridge that formed it will sink, because that plate will be the colder and heavier of the two. In either case, a deep oceanic trench will develop where the sinking plate turns down to start its long dive into the earth's mantle.

Continental crust is too light to sink into the mantle, and remains permanently on the earth's surface. When two continents meet at a collision plate boundary, neither plate can sink. Then one of the two plates must break elsewhere, to form a new plate boundary where lithosphere can sink into the earth's mantle.

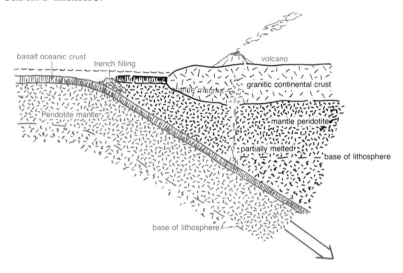

Cross-section of a typical plate collision boundary.

The basalt of the oceanic crust becomes altered and takes on water with time. As it sinks into the earth's mantle it heats up and begins to dehydrate. As it reaches a depth of about 60 miles, large quantities of red hot steam rise from the sinking plate causing melting in the already hot mantle peridotite above. In most cases, the basalt magma and some steam erupt to form a chain of volcanoes, which trends parallel to the trench, and lies about 50 miles away from it. If the basalt magma rises into the lower part of a raft of continental crust, where the rocks are almost hot enough to melt anyway, they melt large volumes of granitic magma. That magma may erupt through volcanoes to form common volcanic rocks such as white rhyolite and gray andesite. We will point those out in the Cascades. However, most of the granitic magma crystallizes at depth to form large masses of new granite, which abound in the mountains all across northern Washington.

After the oceanic crust has lost its water, the slab of relatively cool and rigid peridotite continues to sink into the mantle. Eventually it warms up, and thus loses its identity as relatively cool and rigid mantle peridotite that belongs to the lithosphere. It blends back into the mantle.

New Continental Crust

Most maps and globes clearly show a number of trenches around most of the margin of the Pacific Ocean, but not off the west coast of North America. Nevertheless, an oceanic trench does lie off the coast of the Pacific Northwest from northern California to southern British Columbia. The maps don't show that trench because it is completely stuffed with sediment.

Muddy sediments that accumulate on the ocean floor are too light to sink into the earth's mantle along with the rest of the lithosphere. So the sinking slab jams them rudely into the trench and leaves them there, a chaotic mass of severely deformed sedimentary rocks that originally accumulated over a long period of time and vast expanse of ocean floor. The sinking ocean floor sweeps all those muds and sands together into the trench. Most trench fillings also include large slabs of oceanic crust and peridotite from the mantle that somehow sheared off into the trench, instead of sinking back into the interior of the earth.

Sediments buried deep within the trench get very hot and recrystallize to become what geologists call metamorphic rocks — streaky gneisses and schists which sparkle with big mineral grains, and hardly resemble the original sedimentary rocks. Some of the sediments in the trench get hot enough to melt to form granitic magma, which crystallizes to become granite and related igneous rocks.

Gneiss, schist, and granitic igneous rocks differ considerably in appearance, but they consist of basically the same minerals combined in different proportions and patterns. Those rocks, the stuff of the continental crust, begin their history in the depths of oceanic trenches. That seems strange at first thought, but it does make sense.

An exposure of banded gneiss. It started as layers of muddy sediment in the oceanic trench.
U.S. Geological Survey photo

Sediments that accumulate on the edge of the ocean floor and in the trench consist of material eroded from the continents, and therefore have the same average chemical composition as continental crust. So it is not surprising that metamorphism or melting in hot depths of an oceanic trench converts them back into continental crustal rocks. Oceanic trenches recycle the stuff of the continents, making new continental crust out of the eroded debris of old continental crust. Eventually, the trench filling becomes a new range of mountains.

Trenches exist only because a plate is sinking into the mantle, and they cease to exist when the plate stops sinking. Then all the light sedimentary and metamorphic rocks that had filled the trench rise, just as a marshmallow floats to the surface of a cup of cocoa. If the trench was deeply filled, the rocks that had been stuffed into it rise high, and become a new mountain range. The Olympic Range is a good example of a risen trench filling, as are large parts of the North Cascades, and the area west of Okanogan Valley.

Trenches relentlessly swallow oceanic crust at a rate that probably averages about 2 to 3 inches per year. Although that is much slower than a snail's pace, it is fast by geologic standards. A trench can gobble up more than 32 miles of ocean floor every million years, about 320 miles in ten million years. It adds up. If the sinking plate includes a raft of continental crust, it must eventually arrive at the trench as the last of the oceanic crust between it and the trench disappears. Then the sinking plate can sink no more, because continental crust consists of rocks much too light to sink into the mantle.

If the sinking plate was descending beneath a continent, as most do, then the arrival of a second continent will leave nothing to sink into the trench, and the trench will cease to exist. The continental rocks newly formed deep in the trench filling will then float upward, and suture the two older masses of continental rock together like a welding bead run between them. Meanwhile, a new trench may form elsewhere. That is how the mosiac of pieces that now forms Washington was assembled. But, we have gotten ahead of the story.

THE MISSING PIECE OF NORTH AMERICA

Sometime roughly a billion years ago, give or take two hundred million or so years, a new spreading center formed beneath our continent, and split it along a line that roughly parallels the eastern border of Washington. The Red Sea, a new ocean with an oceanic ridge running down its length, is now splitting Arabia from Africa in exactly the same way. No one knows where our continent was when it split, or what its shape was then. Neither does anyone know where the missing piece went, but many geologists suspect that it now forms a large

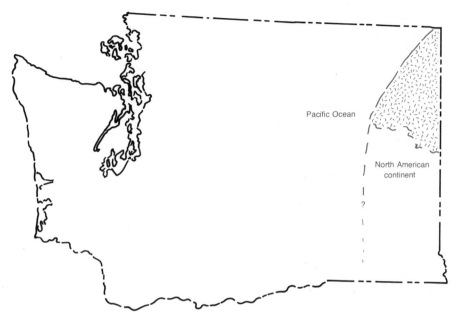

Pacific Ocean

North American
continent

?

This strip of continental crust is all that existed of Washington from a billion years ago until about 100 million years ago. Oceanic trenches assembled the rest of the state during the last 100 million years.

area of eastern Siberia and northern China. Certainly there are rocks in those areas a billion or more years old that closely resemble rocks of the same age in northeastern Washington.

Then, for the next 800 million years or so, the west coast of North America lay within what is now eastern Washington. An ocean spread away to the west just as it does today, but we need not suppose that it was the same ocean we see today. The coast that formed in eastern Washington sometime around a billion years ago remained geologically quiet for about 800 million years. During that time, it accumulated a broad continental shelf and coastal plain, a vast apron of sedimentary rocks along its margin. In many ways, that vanished coastline must have resembled the modern east coast of the United States with its broad coastal plain.

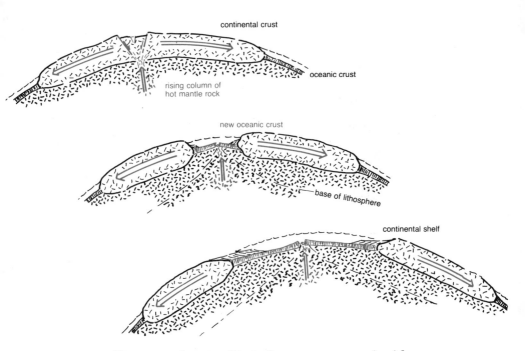

How a continent splits to form a new oceanic ridge.

ASSEMBLING THE PIECES OF WASHINGTON

An Old Ocean, and a New One

As recently as 200 million years ago, North and South America, Europe, and Africa were all part of a vast and temporary supercontinent, which some geologists call "Pangaea." Then, about 200 million years ago, a new oceanic ridge that has since created the Atlantic Ocean split that supercontinent, and the American continents began to move away from Africa and Europe. At first, the Atlantic Ocean was no wider than the Red Sea now is, but, like the Red Sea, it was growing at a rate of about 2 to 3 inches per year. It still grows at that rate as the American continents continue to draw away from Europe and Africa.

As the Atlantic Ocean opened, a new oceanic trench formed along the western coasts of the American continents. Ever since then, the floor of the Pacific Ocean has been slipping through that trench, below the westward moving continents.

So the Pacific Ocean shrinks precisely as the Atlantic Ocean grows, and the North American continent scrapes up everything in its path as it moves west, and as the floor of the Pacific Ocean slides through the trench along its western edge to vanish into the mantle.

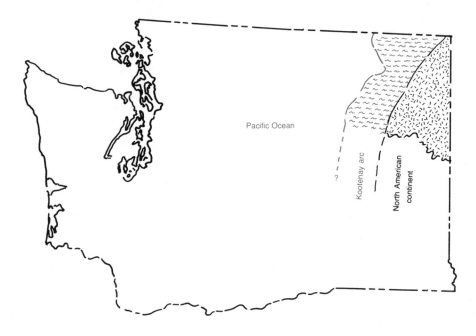

The Kootenay arc in Washington. The dashed line shows the probable continuation of the Kootenay arc beneath younger volcanic rocks south of Spokane.

The old coastal plain that formed along the west coast of North America after the split of a billion or so years ago became the first victim of the collision between North America and the floor of the Pacific Ocean. We see it now crushed into a belt of tightly folded sedimentary rock that traces the former margin of the North American continent. Geologists call the part of that long belt of folded rocks in British Columbia and Washington the Kootenay arc. It disappears beneath younger volcanic rocks west of Spokane. It is difficult to be sure, but the Blue and Wallowa mountains of Oregon and the Sierra Nevada of California may be the southward continuation of the Kootenay arc.

As the new trench gobbled up the floor of the Pacific Ocean, red hot steam and molten basalt rose above the sinking slab of lithosphere. They melted old continental crust in the western margin of the North American continent to create enormous volumes of granitic magma, nearly all of which crystallized below the surface without erupting. Western Montana, Idaho, and northeastern Washington all contain enormous areas of granite and similar rocks formed after the first collision between continent and ocean floor in eastern Washington.

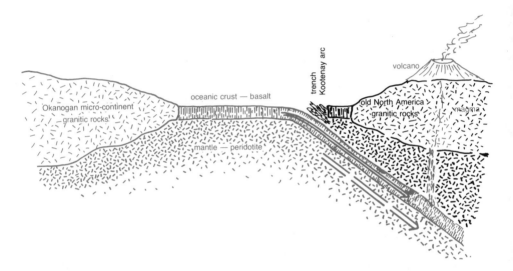

When the floor of the Pacific Ocean began to sink beneath North America, it crumpled the old coastal plain against the edge of the continent to make the Kootenay arc.

Adding the Island Continents

When North America began to move west some 200 million years ago, several small continents existed in the eastern Pacific Ocean. They probably resembled some of the small island continents that still exist in the Pacific, such as Borneo, New Zealand, New Guinea, and Japan. As the ocean floor slipped into the trench, island micro-continents collided with western North America, one after the other. They now form geologically distinctive regions of the Pacific Northwest, British Columbia, and Alaska. Vast expanses of oceanic crust

13

attached to the small continents added their own contribution to the mosaic patchwork that is Washington.

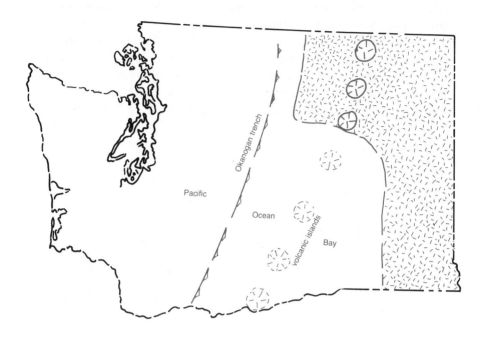

The geography of Washington as it may have looked after the Okanogan micro-continent docked about 100 million years ago.

The first major addition was the Okanogan micro-continent, which joined North America about 100 million years ago. It now forms the highland region between the Columbia and Okanogan rivers in northeastern Washington, and extends far north into British Columbia. When the Okanogan micro-continent landed, the original Kootenay trench in eastern Washington vanished because there was no more ocean floor to sink there.

However, North America was still moving west, and the floor of the Pacific Ocean still had to sink, so a new trench formed off the west coast of the Okanogan subcontinent, in the general area of the present Okanogan Valley. Then a chain of volcanoes formed above the slab of lithosphere sinking into the new Okanogan trench. Rocks erupted from those volcanoes still exist along the crest of the Okanogan subcontinent, and probably continue south under the younger volcanic rocks of the Columbia Plateau. Meanwhile, the North American continent, still moving west, was approaching another island micro-continent in the Pacific.

The North Cascade micro-continent docked about 50 million years ago, and added another large chunk of continental crust to Washington and British Columbia. Its arrival killed the trench that had existed in the area west of the Okanogan Valley, just as the Okanogan micro-continent had eliminated the original Kootenay trench. However, a trench already existed along the west coast of the North Cascade micro-continent before it joined North America, and that trench still swallows the ocean floor off the west coast of Washington.

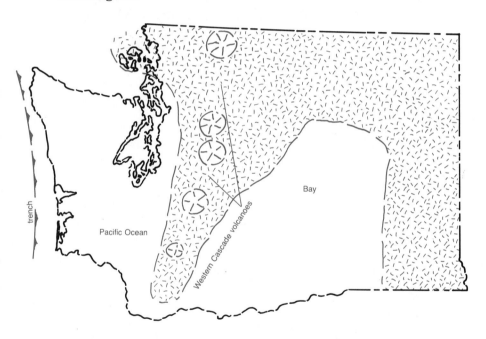

The geography of Washington as it may have looked after the North Cascade micro-continent docked about 50 million years ago.

The Cascades and the Columbia Plateau

Volcanoes had been erupting on the North Cascade micro-continent long before it came within sight of North

America, and continued to erupt after it docked. They covered the southern part of the subcontinent so thoroughly that it is difficult now to trace some of its boundaries.

Starting sometime around 25 million years ago, the old Cascade volcanoes, which geologists call the Western Cascades, snuffed out. Then, there was an intermission in Cascade volcanic activity that lasted for approximately 10 to 15 million years. Meanwhile, the focus of volcanic activity shifted east. A long series of enormous eruptions from several centers in southeastern Washington, several parts of Oregon, and northeastern California built the Columbia Plateau, one of the world's most spectacular volcanic provinces. The Washington portion of the Columbia Plateau consists entirely of black basalt lava flows, almost all erupted from the southeastern corner of Washington and nearby Oregon.

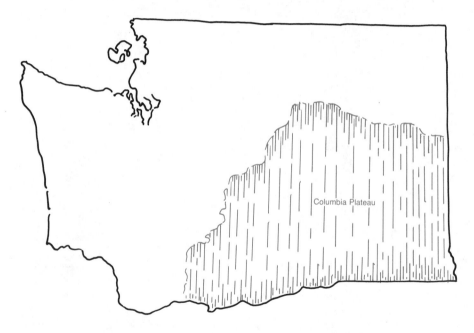

The Washington portion of the Columbia Plateau.

As eruptions ceased in the Columbia Plateau, the focus of volcanic activity shifted east to the Snake River plain of Idaho, and back to the Cascades. Both of those areas are still volcanically active. Renewed activity in the Cascades built the big volcanoes that geologists call the High Cascades. No one knows why volcanic activity in the Columbia Plateau coincided with a lull in the Cascades, but we will offer some conjecture.

Volcanoes like those in the Cascades erupt only parallel to oceanic trenches where one plate slides beneath another, not along transform boundaries where one slides past another. Perhaps a bend in the plate boundary caused the Pacific plate to slide northward past the North American plate for a few million years, instead of sinking beneath it. If the Pacific plate did shear past the North American plate, that motion may have wrenched the area of the Columbia Plateau enough to open fractures, thus causing the basalt magma to melt in the mantle, and erupt as the enormous lava flows that form the Columbia Plateau. Something of the sort may be happening today in the area east of the San Andreas fault of California.

The Willapa Hills, Olympic Peninsula, and Puget Sound lowland

Bedrock beneath the Willapa Hills and the Olympic Peninsula consists entirely of oceanic material. The same is probably true of the Puget Sound lowland.

Everywhere west of the Cascades, bedrock consists mostly of oceanic crust, old Pacific Ocean floor now above sea level. Erosion carved the Willapa Hills of southwestern Washington into a slab of oceanic crust that still lies almost as flat as it formed, even though it now stands about 2 miles higher than most of the world's inventory of oceanic crust. The Olympic Mountains consist of rock that was stuffed into the trench that exists off the coast of Washington. Now that part of the trench filling has floated up, and pokes through the slab of oceanic crust that lies beneath the Willapa Hills, as though it were a fist shoved through a sheet of foam rubber. The same trench filling must continue south beneath the Willapa Hills, and keep that expanse of heavy oceanic crust floated above sea level.

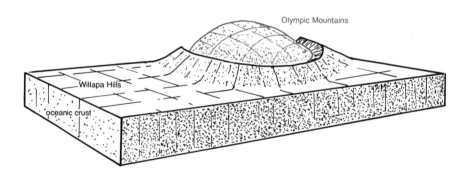

The same slab of oceanic crust that lies almost flat beneath the Willapa Hills rises steeply at the south end of the Olympic Mountains, and wraps around their eastern and northern margins.

The Puget Sound lowland lies between the Olympic Peninsula and northern Willapa Hills on the west and the Cascade subcontinent on the east. It is difficult to say much about the bedrock in the Puget Sound lowland because it lies beneath a thick blanket of glacial debris. Nevertheless, it seems very likely that the bedrock beneath the Puget Sound lowland is also oceanic crust.

THE CLIMATIC RECORD

From the Wet Tropics to the Desert

Ample evidence shows that the Pacific Northwest had a wet tropical climate while the big eruptions were building the Columbia Plateau. And the rocks also show that the climate of the region was extremely dry for about 8 million years after most of the Columbia Plateau formed. However, the evidence of tropical and arid climates is most visible in the Columbia Plateau, so we discuss it in that chapter. The ice age glaciations, on the other hand, left their mark in the landscapes of many parts of Washington.

Ice Ages and the Pleistocene Epoch

Something that no one understands happened to the earth sometime between 2 and 3 million years ago to make the climate extremely unstable. Ever since then, throughout the span that geologists call Pleistocene time, our planet has experienced periodic ice ages. Although perfectly preserved relics of the last ice age surround us in much of our modern landscape, no one seems to know what caused the ice ages, how many there were, or when the next one may start. Neither do we know much about ice age climates, except that they were wet and cold enough to maintain large glaciers.

Many ice ages came and went, and it seems reasonable to assume that they all affected Washington. However, direct evidence of the earlier ice ages is hard to find because big glaciers tend to bury the record of their predecessors, if they do not bulldoze it off the landscape. But we do know that large glaciers covered most of the northern third of Washington, as well as all the higher mountains, during the last two ice ages.

Extent of Ice Cover

Melting glacial ice deposits a distinctive sediment that geologists call till. It consists of rock fragments of all sizes from clay to boulders mixed indiscriminately together. Glaciers dump till around their margins to form ridges called moraines,

A boulder in a roadcut in glacial till beside Washington 20 about 3 miles east of Tonasket, where the melting ice of the Okanogan lobe left it approximately 12,000 years ago. Like many such boulders, this one has polished surfaces covered with fine scratches. Nothing but glacial transport can leave a boulder so marked.

Most of the northern third of Washington lay beneath a nearly continuous sheet of glacial ice during the maximum of ice age glaciation. This map does not show the glaciers that filled the high mountain valleys of the Cascades.

which neatly preserve the outline of the ice that deposited them. Meanwhile, torrential flows of summer meltwater sweep great loads of sediment off the ice, and dump it in deposits called outwash, which typically consists of beds of clean sand and gravel laid down below the moraines. So the glaciers left the archives of the ice ages in their legacy of morainal ridges with vast expanses of outwash spreading away from them.

LOOKING AHEAD

History, it is said, is merely the prelude to the present. If so, the present must likewise set the stage for the future, in geologic as well as in human events. If current geologic trends continue, we can imagine how the Pacific Northwest may change during the next several tens of millions of years.

The Doomed Oceanic Ridges

The Gorda and Juan de Fuca oceanic ridges a few hundred miles offshore now generate new oceanic crust, which sinks into a trench immediately off the modern coast. The short lengths of ridge are remnants of an oceanic ridge that roughly paralleled the entire west coast of North American until sometime around 15 or 20 million years ago. The Farallon plate, moved away from that oceanic ridge and into a trench that also followed the entire west coast. Then the last of the Farallon plate began to vanish into the trench as continuing westward movement of North America brought the trench and ridge together off the coasts of California and British Columbia. One remnant of the original oceanic ridge and trench, and of the intervening Farallon plate, survives off the Pacific Northwest.

Disappearance of the Farallon plate along most of the west coast of North America joined the North American and Pacific plates along the line where the oceanic ridge and trench met. The Pacific plate now slides north along the notorious San Andreas fault of California, and the similar, but less notorious, Queen Charlotte fault off the coast of British Columbia. Both faults formed and began to move as the last of the Farallon plate went down the trench. They are transform plate boundaries.

The plate tectonic situation in the Pacific Northwest. The oceanic ridges generate new oceanic crust, shown in the stippled area, which is actually a small remnant of the old Farallon plate. That small plate sinks through the trench and beneath the North American continent. North and south of the oceanic ridge, the Pacific plate slips north past the North American plate on the Queen Charlotte and San Andreas faults.

Sometime within the next 10 to 15 million years, that last remnant of the Farallon plate off the Pacific Northwest will slide into the trench, and the ridge and trench will meet along the entire west coast. That will connect the San Andreas and Queen Charlotte faults. Then, the Pacific plate will move north to the Aleutian trench along a single fault, a transform plate boundary, that will follow the west coast of North America from Mexico to Alaska.

Final disappearance of the small plate off the coast of the Pacific Northwest will kill the trench that now exists along the coast of Washington. When that happens, the Cascade volcanoes will flicker out for lack of fuel. They depend for their existence upon molten basalt magma and steam rising above the sinking plate. Disappearance of the trench will also permit the trench fillings which now extends east beneath the Willapa Hills, to float, and become a new range of mountains along the entire western coasts of Washington and Oregon. The Olympic Mountains are merely the first part of the range to emerge from the depths where the rocks were scraped together.

A Future Micro-continent

A large slice of California west of the San Andreas fault, which geologists call the Salinian block will eventually detach from California and move north with the Pacific plate as a long, narrow micro-continent. It will pass along the coast of the Pacific Northwest at the stately rate of about 2 inches per year, and eventually dock onto southern Alaska when it arrives at the Aleutian trench.

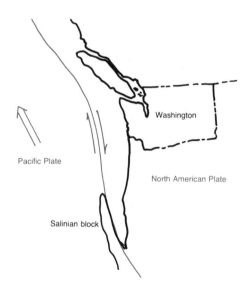

The Pacific Northwest as it will be after the last of the Farallon plate has vanished into the mantle, and the Queen Charlotte and San Andreas faults have joined to make a continuous transform boundary between the North American and Pacific plates. A slice of California is beginning to move out into the Pacific as a new micro-continent.

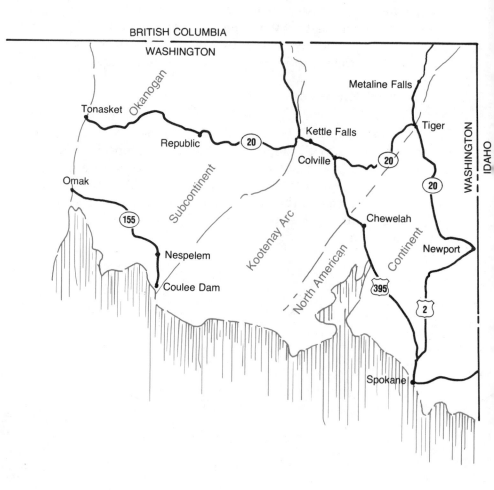

The three major geologic provinces of northeastern Washington.

II
THE NORTHEASTERN HIGHLANDS

The highlands of northeastern Washington include three major geologic provinces. All contain folded sedimentary rocks and large quantities of granite, which looks about the same everywhere. So the rocks exposed along the roads in the different provinces are not really as distinctive as you might expect. Nevertheless, the distinction into separate provinces is necessary if we are to make sense of the rocks.

We will consider the three provinces in the order of their becoming part of Washington, which is also their order from east to west. The oldest is the western edge of the North American continent, a small patch of the Northern Rocky Mountains in the northeastern corner of Washington. Next is the old coastal plain and continental shelf that once lay along the western margin of North America, and now survives as a belt of tightly folded sedimentary rocks full of granite intrusions, the Kootenay arc. Finally, the Okanogan highlands west of the Kootenay arc are the southern end of the Okanogan subcontinent. The southern edges of all three geologic provinces vanish beneath the Columbia Plateau.

THE OLD NORTH AMERICAN CONTINENT

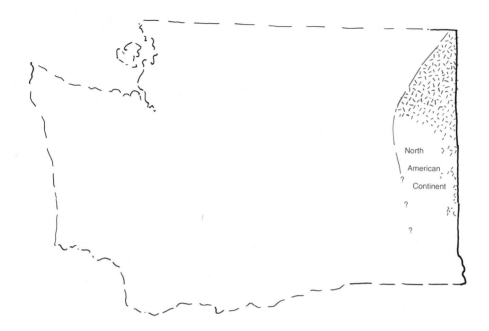

Northeastern Washington is part of the old North American continent, which extends south beneath an eastern strip of the state. The southern part of the map dissolves into question marks because younger volcanic rocks cover the older rocks.

Sometime roughly a billion years ago, give or take two hundred million or so years, a new spreading center formed beneath our continent, and split it along a line that trends from north to south through eastern Washington. No one knows certainly where the detached piece of the old continent went, but we can be sure that it still exists somewhere, because continental crust floats permanently on the earth's surface.

The northeastern corner of Washington belongs to the old North American continent, and contains rocks like those in nearby areas of the Northern Rocky Mountains. The oldest rocks are ancient continental crust, which consists of a complex of granite, gneiss and schist, all considerably more than two billion years old. In large areas, the ancient continental crust lies beneath a thick cover of sedimentary rocks deposited sometime around one billion years ago, the Precambrian rocks of the Belt supergroup. Large masses of much younger granite intrude both the old continental crust and the ancient sedimentary rocks. They formed as steam and molten basalt rose above the slab of ocean floor that sank into a trench in eastern Washington, and melted the lower part of the old continental crust.

THE KOOTENAY ARC

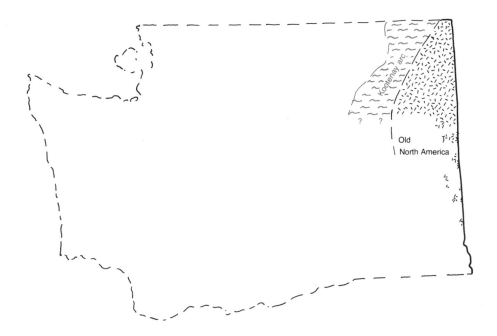

The Kootenay arc borders the North American continent, and then disappears beneath basalt lava flows of the Columbia Plateau west of Spokane. The dashed lines indicate its probable southward extension.

For approximately 800 million years after the continent split, the coastline in eastern Washington was a quiet continental margin where layers of sediment accumulated, as now happens along the east coast of North America. By about 200 million years ago, the old continental margin had acquired a broad coastal plain and continental shelf. About that time another continental split separated North America from Europe and Africa along a line that is now the middle of the

27

Atlantic Ocean. The North American continent began to move west over the floor of the Pacific Ocean, which sank into an oceanic trench along the western edge of the continent, in the area that is now eastern Washington.

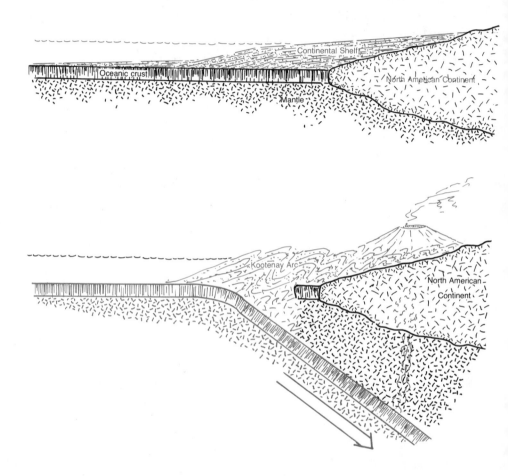

The collision between continent and the floor of the Pacific Ocean jammed the old continental shelf against the continent to form the Kootenay arc. Sedimentary rocks melted deep in the trench to form granite magma.

The continental shelf had accumulated mostly on oceanic crust while it was firmly attached to the continent. When that oceanic crust began to slide into the trench, it crushed the continental shelf into the edge of the continent, and telescoped the layered sedimentary rocks into the tight folds we see in the Kootenay arc. The effect was as though the sedimentary rocks had been deposited on an escalator that suddenly began to move.

It is difficult to imagine that the Kootenay arc was once a broad expanse of coastal plain and continental shelf because it is now a relatively narrow band of eastern Washington, less that an hour's drive from one side to the other. But notice that many of the layers of rock in that area stand nearly vertically. If you could pull those tight folds out flat, the Kootenay arc would stretch like an accordion into quite a respectable coastal plain.

THE OKANOGAN SUBCONTINENT

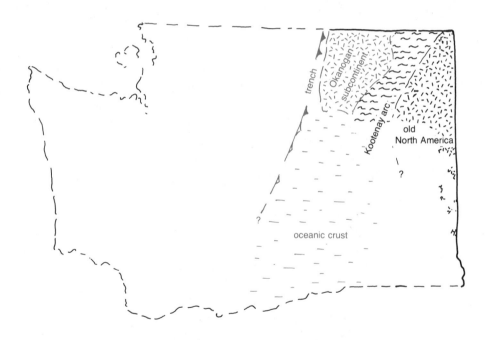

The Okanogan subcontinent comprises a large part of northeastern Washington. Oceanic crust that landed with the subcontinent probably lies beneath a large part of south-central Washington, where younger volcanic rocks of the Columbia Plateau cover it completely.

The Okanogan micro-continent was an island about the size of California until the same trench that crushed the Kootenay arc into folds consumed all the ocean floor that separated it from North America. That landed the micro-continent against the folded sedimentary rocks of the Kootenay arc to become the Okanogan subcontinent we now know. Geologists recognized the Okanogan highlands as a distinctive region of Washington decades before anyone realized that they were once a separate small continent.

It is hard to be exactly sure when the Okanogan micro-continent docked, but a date sometime around 100 million years ago seems likely. North America began moving west over the floor of the Pacific Ocean about 200 million years ago, so the Okanogan micro-continent could have come several thousand miles during the approximately 100 million years that elapsed before it joined North America.

The Columbia River now traces the eastern and southern margins of the Okanogan subcontinent in Washington, the Okanogan River its western margin. The subcontinent extends hundreds of miles north into British Columbia. No one knows how far south the Okanogan subcontinent may extend, because its southern edge disappears under the younger basalt lava flows of the Columbia Plateau. However, there is no reason to assume that it continues very far beyond where we last see it.

The trench that reeled the Okanogan micro-continent into North America destroyed itself as it swallowed the last of the ocean floor that had separated the two continental rafts. Elimination of that trench saved a large expanse of oceanic crust that extended south from the Okanogan micro-continent from sinking into the mantle, the normal destiny of old ocean floor. That tract of oceanic crust must still lie beneath part of eastern Washington, where younger rocks bury it completely.

The Okanogan Trench

But North America was still moving west, and something had to give, even though the old trench no longer existed. So the plate broke along the west side of the newly docked Okanogan subcontinent to form a new trench, which followed the line of the Okanogan Valley through north central Washington and central British Columbia. It is difficult now to trace that trench south from the Okanogan Valley because younger rocks cover it. However, it probably angled southwest through Washington toward western Oregon.

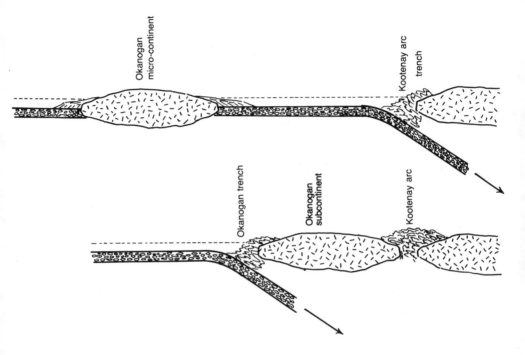

As the Okanogan subcontinent became part of North America, the trench that formed the Kootenay arc stopped, and a new one formed on the west side of the Okanogan Valley.

Oceanic crust promptly began to sink below the west coast of the newly attached subcontinent, leaving its burden of sedimentary rocks in the new Okanogan trench. That trench filling now forms much of the bedrock in and west of the Okanogan Valley. Steam and melted basalt rose above the sinking slab of lithosphere, and melted large volumes of the newly docked Okanogan subcontinent. A line of volcanoes formed along the crest of the newly attached subcontinent, and probably extended south through central Washington and into Oregon. Great masses of granitic magma invaded the older rocks of the Okanogan subcontinent, and crystallized at depth to form younger granite plutons.

31

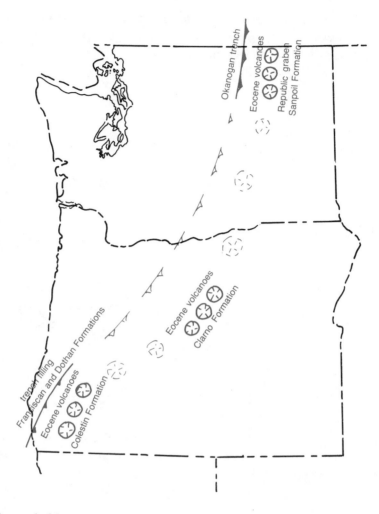

The probable course of the Okanogan trench and its associated volcanic chain, now mostly buried beneath younger volcanic rocks in the interval between central Washington and northern California.

A Volcanic Chain and the Republic Graben

Faults that trend generally north and south dropped long slices up the center of the Okanogan subcontinent. Geologists refer to those dropped slices as the Republic graben. They divide the subcontinent into three parts, the Kettle highlands east of the Republic graben, the Republic graben itself, and the Okanogan Valley. Bedrock in both the Kettle and Okanogan highlands is a complex of old continental crust and younger granite. That in the Republic graben is volcanic rock erupted during Eocene time, mostly gray andesite and white rhyolite. The graben, its filling of volcanic rocks, and the trench offshore all fit together into a logical pattern.

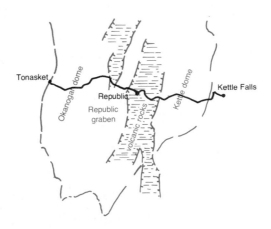

The Republic graben splits the Okanogan subcontinent into three major segments.

Volcanic chains typically form parallel to oceanic trenches, and about 50 miles from them. Therefore, it is no surprise to find the volcanic rocks in the Republic graben following a north to south trend about 50 miles inland from the old Okanogan trench. Like everything else in northeastern Washington, the volcanic rocks in the Republic graben disappear beneath the thick cover of basalt in the Columbia Plateau. Although nothing is more opaque than thousands of feet of basalt, there are clues that encourage us to trace the volcanic chain southward.

Two deep and unsuccessful wildcat oil wells drilled near Yakima penetrated thick sections of Eocene volcanic rocks, the same age as those in the Republic graben. That area is generally southwest of the Republic graben, and east of our projected trend of the Okanogan trench. Following the same trend farther south leads to the Ochoco Mountains of Oregon, which consist of Eocene volcanic rocks, the Clarno formation.

They also lie east of the projected southward continuation of the Okanogan trench.

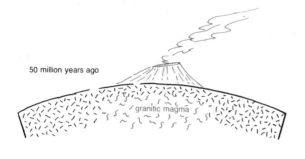

50 million years ago

granitic magma

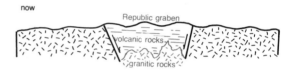

now

Republic graben

volcanic rocks

granitic rocks

Formation of the Republic graben.

Active volcanic chains typically stand on the crest of a broad arch in the earth's crust. Travelers passing through the Columbia Gorge can see that the modern High Cascades stand on such an arch, because the layered basalt lava flows of the Columbia Plateau tilt down both east and west of the Cascades. It seems likely that those broad arches form beneath active volcanic chains simply because large volumes of hot rock at depth cause the crust to expand. If so, then the arch should collapse after the volcanoes become extinct and the hot rocks beneath them cool. It is easy to understand the Republic graben as simply the dropped keystone of a swollen crustal arch that expanded when volcanoes were active in the area, and collapsed after they became extinct.

The Kettle and Okanogan Domes

Not all the molten granitic magma erupted to become the rhyolite and andesite that fills most of the Republic graben.

Large masses crystallized below the surface to form granite plutons and batholiths that intrude the older rocks of the Kettle and Okanogan highlands. Enormous volumes of old rocks deep in the crust of the Okanogan subcontinent became so hot that they melted, and then floated upward through the heavier metamorphic rocks above them where they crystallized to form great intrusions of young granite.

The Kettle dome occupies a large area in the Kettle highlands east of the Republic graben, the Okanogan dome and even larger area in the Okanogan highlands to the west. Each contains in its core an enormous mass of granite that pushed up as molten magma through the older and cooler rocks above, dragging them along. Now we see the older metamorphic rocks surrounding the granite on all sides, wrapping around the igneous core of the dome.

U.S. 2, Washington 20, 31
Spokane—Newport—Canadian Border
122 mi. 197 km.

The area between Spokane and Newport is near the western edge of the old North American continent. Rocks exposed along and near the road include billion year old sedimentary rocks of the Belt supergroup, even more ancient continental crustal rocks that lie beneath those ancient sedimentary rocks, and masses of granite that intruded as molten magma during Cretaceous time. Some of the flood basalt flows of the Columbia Plateau lapped north into the valleys.

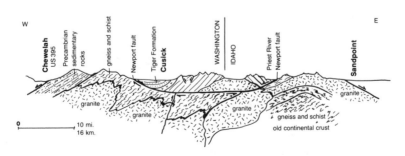

Section across the line of Washington 20 near Cusick.

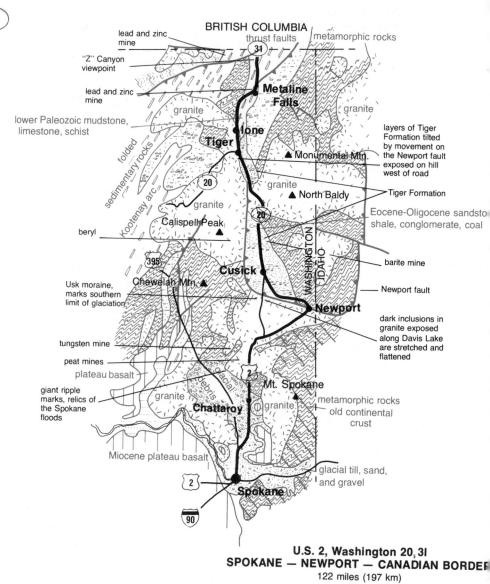

BRITISH COLUMBIA

thrust faults · metamorphic rocks

lead and zinc mine

"Z" Canyon viewpoint

lead and zinc mine

lower Paleozoic mudstone, limestone, schist

granite

Metaline Falls

granite

Ione

Tiger

layers of Tiger Formation tilted by movement on the Newport fault exposed on hill west of road

▲ Monumental Mtn.

folded sedimentary rocks

Kootenay arc

granite

granite

▲ North Baldy

Tiger Formation

Eocene-Oligocene sandstone, shale, conglomerate, coal

beryl

Calispell Peak ▲

barite mine

Usk moraine, marks southern limit of glaciation

Cusick

WASHINGTON IDAHO

Newport fault

Chewelah Mtn. ▲

Newport

tungsten mine

dark inclusions in granite exposed along Davis Lake are stretched and flattened

peat mines

plateau basalt

giant ripple marks, relics of the Spokane floods

granite

glacial debris

Mt. Spokane

Chattaroy

granite

metamorphic rocks old continental crust

Miocene plateau basalt

glacial till, sand, and gravel

Spokane

U.S. 2, Washington 20, 31
SPOKANE — NEWPORT — CANADIAN BORDER
122 miles (197 km)

0   30 mi.
           50 km.

Glaciation

The highway follows the valley of the Little Spokane River north out of Spokane across a broad valley with deep deposits of glacial debris in its floor. Glacial ice advanced only about as far south as Newport in the eastern edge of Washington. There it impounded the stream drainage to create Glacial Lake Spokane, which flooded the valleys in this area. The glacial debris consists partly of sediment deposited in the glacial lake, partly of outwash carried in as the ice melted.

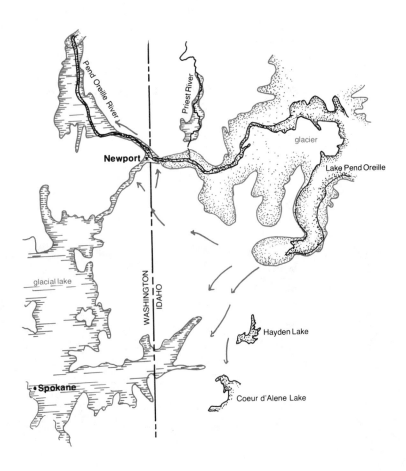

The general distribution of glacial features in the area north of Spokane at the maximum of the last ice age.

Plateau Basalt

Flat-topped hills several miles east of Colbert and Chattaroy are basalt lava flows of the Columbia Plateau, which flooded into this broad valley. Evidently, the valley already existed during late Miocene time when the flood basalts erupted. Wind blown dust exactly like that in the Palouse Hills south of Spokane caps the basalt flows in this area at the northern limit of both plateau basalt and loess.

The flanks of those flat-topped basalt hills contain numerous exposures of the Latah formation, white sedimentary rocks deposited in lakes impounded behind the big lava flows. Clay pits have opened several good exposures near Chattaroy. Old clay pits in the Latah formation are good places to look for layers of white volcanic ash full of beautifully preserved fossil leaves. Most of the leaves are of species similar to those that live today in southeastern United States and the Caribbean region. They clearly show that the Pacific Northwest had a warm and wet climate during late Miocene time.

Granite

An exposure of granodiorite in a roadcut just south of the Pend Oreille River. The dark inclusions are fragments of older rock that the magma picked up, but did not assimilate before it crystallized.

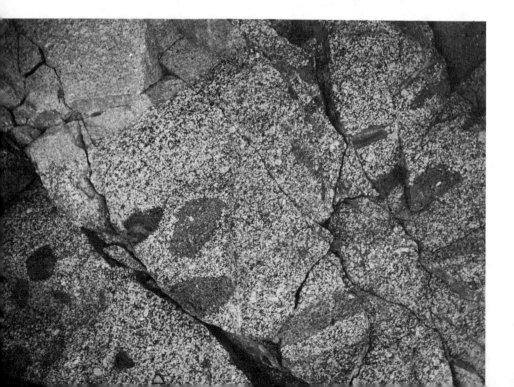

Higher hills both east and west of the road in the area near Spokane are outposts of the Northern Rocky Mountains. Most of those hills consist of granite and granodiorite that crystallized during Cretaceous time. These granitic rocks are closely related to the very large mass of granite, called the Idaho batholith, that occupies most of central Idaho, and extends into western Montana. Mount Spokane State Park, the prominent high peak east of Chattaroy, is near the center of the Spokane dome, a large mass of younger granite that intruded the ancient continental crustal rocks during late Cretaceous time, about 100 million years ago. Exposures of Belt sedimentary rocks appear east of the park.

As sometimes happens, the granite magma brought uranium with it. Numerous uranium prospects, and several old uranium mines, lie in a broad belt that passes from north to south through the Mount Spokane area. The only district in Washington that has produced more uranium is in the hills west of Ford, about 40 miles west of this highway.

The Old North American Continent

Unfortunately, this route along the eastern fringe of Washington crosses very little of the ancient igneous and metamorphic basement rock of the old North American continent. Large areas of that rock do exist in this part of Washington, but exposures along the road are few and poor. They consist of an incredibly complex assemblage of metamorphic rocks, mostly streaky looking gneisses, along with some granite.

There are some sedimentary rocks along the road, as well as some younger granite. Watch for exposures of gray granite in roadcuts near Ruby. It is easy to recognize the sedimentary rocks because they are layered. Most are ancient formations of the Belt supergroup laid down on top of the old North American continent about a billion or so years ago. Similar rocks cover much of northern Idaho, western Montana, and southeastern British Columbia. Their western limit near the line of the Columbia River, a few miles west of Washington 31, helps define the edge of the old North American continent.

The Newport Fault

Newport lies right at the southern tip of the Newport fault, one of the strangest geologic structures in Washington. On the map, the Newport fault looks like a giant horseshoe lying with its open end facing north. The state line slices the horseshoe neatly down the middle, giving the western half to Washington.

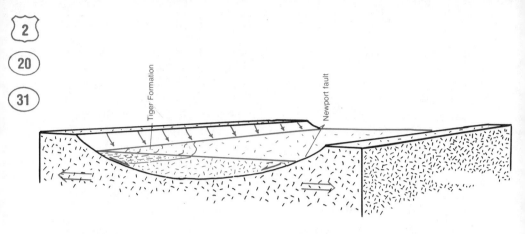

Schematic cross-section showing how the Newport fault moved as the continental crust beneath it stretched.

The open northern end of the Newport fault did not break or move. However, the entire mass of rock enclosed within the loop of the fault dropped relative to the rocks outside it. Evidently, something was pulling the earth's crust in this area, and the spoon-shaped flap within the Newport fault settled as the rocks beneath it stretched.

A thick sequence of slightly tilted sandstones and pebble conglomerates that geologists call the Tiger formation, appears in the area near Tiger. These sediments accumulated during Eocene time, in round numbers approximately 50 million years ago. The Tiger formation exists only within the western part of the Newport fault hairpin, and appears to have filled the basin that formed there as the area within the fault dropped.

Some Glacial Features

Newport nestles in the valley of the Pend Oreille River, which flows north a short distance into British Columbia, and then loops back south to join the Columbia right on the border, just south of Trail.

When the glaciers of the last ice age were at their maximum advance, the ice in Washington reached the area near Newport. In Idaho, it flowed south through the Purcell Valley all the way to the north side of Coeur d'Alene Lake, and then hooked west almost to Spokane. Then the glaciers began to melt back as the last ice age ended. As the shrinking glaciers uncovered the valleys of the Priest River in Idaho, and the Pend Oreille River in Washington, meltwater flooded the area between the ice in the north, and the drainage divide in the south. A lake that drained to the south existed for several hundred years until the ice finally retreated far enough to permit the normal northward drainage to resume.

That ice-dammed lake left a record of itself in large deposits of lake sediment, thinly layered beds of clay, silt, and sand, that cover a considerable area of the valley floor in the Newport area. In some places water wells have penetrated more than 200 feet of glacial lake deposits. The lake lasted until continued retreat of the ice finally permitted the water to empty through the Columbia River.

View north up the canyon of the Pend Oreille River north of Metaline Falls.

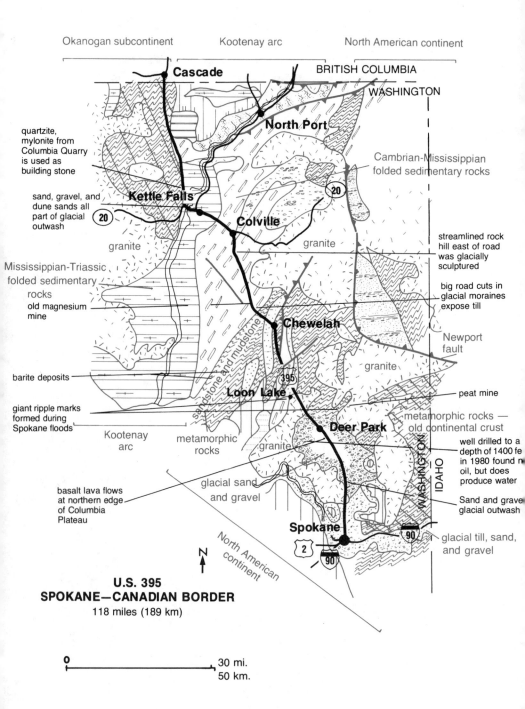

Okanogan subcontinent   Kootenay arc   North American continent

Cascade

BRITISH COLUMBIA

WASHINGTON

North Port

quartzite, mylonite from Columbia Quarry is used as building stone

Cambrian-Mississippian folded sedimentary rocks

sand, gravel, and dune sands all part of glacial outwash

Kettle Falls

20

Colville

granite

granite

streamlined rock hill east of road was glacially sculptured

Mississippian-Triassic folded sedimentary rocks

big road cuts in glacial moraines expose till

old magnesium mine

Chewelah

Newport fault

granite

barite deposits

395

peat mine

Loon Lake

giant ripple marks formed during Spokane floods

metamorphic rocks — old continental crust

Kootenay arc

Deer Park

metamorphic rocks

well drilled to a depth of 1400 fe in 1980 found n oil, but does produce water

granite

basalt lava flows at northern edge of Columbia Plateau

glacial sand and gravel

Sand and grave glacial outwash

Spokane

WASHINGTON

IDAHO

2

90

90

glacial till, sand, and gravel

N

North American continent

U.S. 395
SPOKANE—CANADIAN BORDER
118 miles (189 km)

0   30 mi.
50 km.

U.S. 395
Spokane — Canadian Border
118 mi. 189 km.

The route crosses the western edge of the North American continent between Spokane and Chewelah, the Kootenay arc between Chewelah and Kettle Falls. Between Kettle Falls and the border, the road closely traces the boundary between the North Cascade subcontinent and the Kootenay arc. Much of the area lay beneath glacial ice during the last ice age.

Glaciation

Deposits of glacial sediments, glacial meltwater sediments, and debris left by the Spokane floods cover many of the low areas along most of the route. The thin southern edges of the ice age glaciers fingered slowly into the more southern valleys, and lay gently on the land, leaving little erosional record of their passage. So their exact extent is difficult to determine, and not precisely known except locally.

Except for the highest hills, which stood like islands above the glacier, the country north of Loon Lake lay beneath a blanket of ice, which thickened northward. Loon Lake lies immediately south of the glaciated area, and exists because a glacial moraine at the north side of the lake impounds drainage from the unglaciated bedrock hills farther south. Deer Lake similarly lies behind a glacial moraine dam at its western edge, and floods a valley that lay just beyond the reach of the ice.

Nearly all the hills surrounding Loon Lake are granite, the only exception being a small area of sedimentary rocks on the east side. Precambrian sedimentary rocks about a billion years old surround most of Deer Lake, although granite is exposed along short stretches of its eastern and northern edges. The granite around Loon Lake and part of Deer Lake intruded the Precambrian sedimentary rocks about 100 million years ago.

Glacial ice filled this valley near Colville during the last ice age, and probably covered most of the hills too. A deep fill of glacial sediment deposited as the ice melted now lies beneath the broad valley floor.

The North American Continent

Between Spokane and Chewelah, the road crosses the western edge of the North American continent. Bedrock consists mostly of Precambrian sedimentary rocks belonging to the Belt supergroup, and intrusive masses of younger granite. Older continental basement rock also exists in this region, but not along the highway.

The several formations of Precambrian sedimentary rock are all conspicuously layered, and look quite different to a geologist, although perhaps not to people less fascinated by rocks. Most of the Precambrian sedimentary rocks had already formed when the large piece of western North America drifted off about a billion years ago.

Granite magma intruded the Precambrian sedimentary rocks in this area during three major intervals about 190, 90, and 50 million years ago. All the granitic rocks look about the same regardless of their ages, black speckled gray rocks with no sign of layering. They consist mostly of milky white or pink feldspar, glassy gray looking quartz, and a black mica or hornblende, all in grains large enough to see without a magnifier.

Ore Deposits

As so commonly happens, the intruding granite magmas created a variety of mineral deposits. Several mines in the hills surrounding Chewelah produced lead, silver, copper, and gold, not necessarily in that order of value, during the decades between the 1890's and the 1950's, sporadically since then. Some of the best mines were along the west side of Eagle Mountain, about 3 miles northeast of Chewelah. Most of the ore bodies are quartz and calcite veins in the Precambrian sedimentary rocks near the contacts of granite intrusions. Blue Grouse Mountain, east of Deer Lake, contains several tungsten prospects, which may someday develop into mines.

Several mines in the area southwest of Chewelah produced magnesite between 1916 and 1949. Magnesite is magnesium carbonate, and was in those days the main raw material for magnesium metal. However, this country began to extract magnesium from sea water along the Gulf Coast of Texas during the second World War, and now obtains its entire domestic supply there. Mineral magnesite can not compete commercially with sea water as a source of magnesium; so the mines are closed now, with little prospect of ever resuming operation.

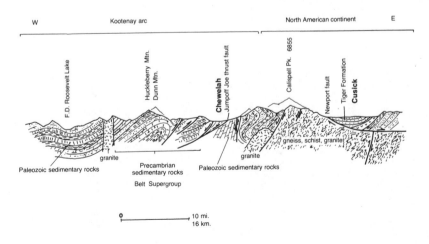

Section Across the line of U.S. 395 near Chewelah.

The Kootenay Arc

Between Chewelah and Kettle Falls, the road crosses the Kootenay arc, once the coastal plain and continental shelf, now a belt of tightly folded Precambrian and Paleozoic sedimentary rocks. The Kootenay arc also contains masses of granite that intruded as molten magma after the sedimentary rocks were already tightly folded. This southern part of the Kootenay arc contains so many granite intrusions that they nearly consume the fold belt, and make it hard to see on the geologic map.

The Continental Suture

Kettle Falls gets its name from a waterfall now drowned beneath the Franklin D. Roosevelt Lake. From that area north, the highway follows a line very close to the eastern margin of the Okanogan subcontinent. Rocks along the road and in the hills west of the road are massive granites and ancient streaky gneisses that belong to the Okanogan subcontinent. Rocks in the hills east of the road are tightly folded sedimentary rocks and granites that belong to the Kootenay arc. An ocean formerly separated the rocks along the road north of Kettle Falls from those in the hills visible in the eastern distance.

A close view of a strongly sheared gneiss exposed in roadcuts beside U.S. 395 just north of its junction with U.S. 2. This is part of the shear zone that formed on the east side of the Kettle dome as the granite core rose. Notice the curving tail on the white feldspar crystal in the right edge of the picture. It shows that the mineral grain rolled clockwise as the rocks above it moved to the right, in this case to the east.

Washington 20
Tonasket—Colville—Tiger
130 mi. 210 km.

Coast to Coast

The short drive between Tonasket and Kettle Falls takes the traveler from one coast of the Okanogan subcontinent, across its highlands and to the opposite coast. In a geologic sense, it is almost the equivalent of driving from one coast of Japan, a modern micro-continent, to the other.

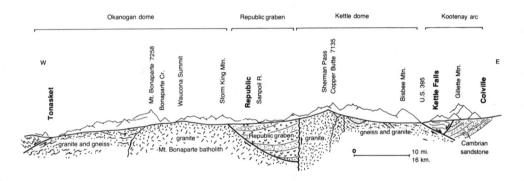

Section along the line of Washington 20 between Tonasket and Tiger.

Kettle Falls is near the east coast of the Okanogan micro-continent. If you stand in Kettle Falls and look east, you gaze across the seam that marks the position of a vanished ocean basin toward the western margin of old North America. Tonasket is on the west coast of that former island continent, near its southern end. And the view west from Tonasket looks across the Okanogan Valley to the highlands of the North Cascade Range. That is the North Cascade subcontinent, once an independent micro-continent far out in the Pacific, now also embedded in western North America.

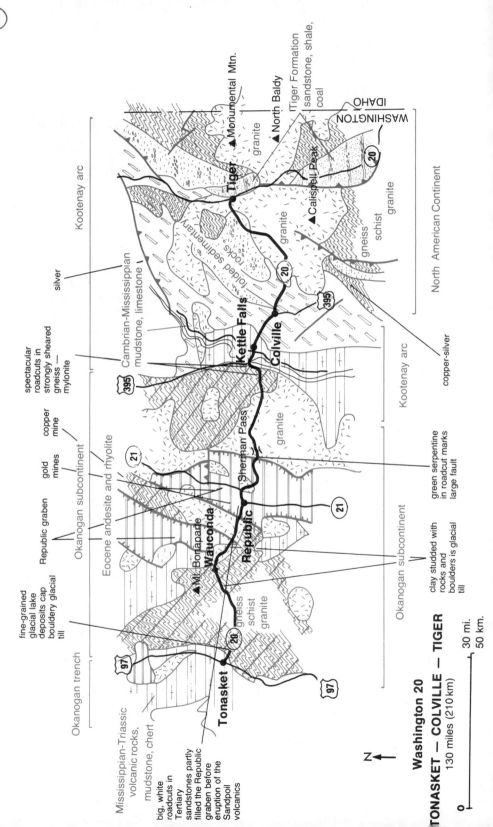

Washington 20
TONASKET — COLVILLE — TIGER
130 miles (210 km)

0 _____ 30 mi.
_____ 50 km.

N

Okanogan trench

Mississippian-Triassic
volcanic rocks,
mudstone, chert

big, white
roadcuts in
Tertiary
sandstones partly
filled the Republic
graben before
eruption of the
Sandpoil
volcanics

fine-grained
glacial lake
deposits cap
bouldery glacial
till

Republic graben

Okanogan subcontinent

Eocene andesite and rhyolite

gold
mines

copper
mine

spectacular
roadcuts in
strongly sheared
gneiss —
mylonite

Cambrian-Mississippian
mudstone, limestone

silver

Kootenay arc

folded sedimentary rocks

clay studded with
rocks and
boulders is glacial
till

Okanogan subcontinent

green serpentine
in roadcut marks
large fault

copper-silver

Kootenay arc

North American Continent

Tonasket

▲Mt. Bonaparte

Wauconda

Republic

Sherman Pass

Kettle Falls

Colville

granite

granite

gneiss
schist
granite

gneiss
schist
granite

▲Calispell Peak

Tiger

granite

granite

▲Monumental Mtn.

▲North Baldy

Tiger Formation
sandstone, shale,
coal

WASHINGTON
IDAHO

97

97

20

21

21

395

395

20

20

20

An exposure of glacial till in a roadcut through a moraine about 6 miles east of Republic. Till is easy to recognize as an unlayered deposit of crudely mixed material of all sizes. It looks as though a bulldozer might have scraped it up.

The rocks in the Okanogan Valley are truly a mess, mostly buried under deep deposits of glacial debris. So far as one can see, bedrock in the Okanogan Valley consists of oceanic material that was stuffed into the Okanogan trench, and then intruded by large masses of molten granite. Those rocks are all that remains of the ocean floor that separated the Okanogan and North Cascade micro-continents until the Okanogan trench gobbled it up.

Okanogan Dome

Most of the route between Tonasket and Republic crosses the Okanogan dome, a mass of granite about 20 miles in diameter completely shrouded in a mantle of gneiss between 5 and 10 miles wide. To any but the loving eye of a geologist, the gneiss looks very much like the granite, except in being streaky. Both the granite and the gneiss appear to have formed as the older rocks of the micro-continent heated, partially melted, and became generally mobile. The rising mass of granite dragged along its mantle of older gneiss.

Radioactive age dates on both granites and gneisses between Tonasket and Wauconda give times of intrusion which we interpret as being between 50 and 70 million years ago. Therefore, the gneiss dome formed long after the Okanogan micro-continent docked. It seems reasonable to relate its formation to the slab of ocean floor that was sinking through the Okanogan trench until 50 million years ago. Basalt magma and steam rising from the sinking slab melted rocks deep in the continental crust to create granite magma, which rose dragging hot, but unmelted, gneiss with it.

The Republic Graben

Between Wauconda and the area just east of Republic, the road crosses the Republic graben. Rocks in this area are andesite and rhyolite, which erupted from volcanoes about 50 million years ago. The best exposures along the highway are in large roadcuts of glaring white rhyolite ash near the town of Republic. The volcanic rocks in the Republic graben have the same chemical composition as some of the granites nearby, and formed at the same time. So, it seems reasonable to conclude that the granite and rhyolite along this road are closely related rocks. They differ mainly in that the rhyolite erupted and the granite did not.

Those Eocene volcanoes should continue south from the Republic graben to connect with the Eocene volcanic rocks in the Ochoco Mountains of central Oregon. Unfortunately, the much younger basalt lava flows of the Columbia Plateau cover everything between. We expect any future deep wells drilled through the Columbia Plateau in the area south of the Republic graben through Ellensburg, Yakima, Toppenish, and Goldendale to encounter buried Eocene volcanic rocks.

Precious Metals

Prospectors discovered gold and silver near Republic just before the turn of the century, and the district has since produced more gold than any other in the state. The High Knob Mine, Washington's most productive and most enduring gold and silver mine, has worked an ore body several miles north of Republic since 1910. The ore is in large quartz veins, and averages about 5 times as much silver as gold, although the gold is far more valuable.

Kettle Dome

Between Republic and Kettle Falls, the highway crosses the Kettle dome, another large mass of intrusive granite enclosed within a mantle of gneiss. It is generally similar to the Okanogan dome, also 50 to 70 million years old, and also formed after the Okanogan micro-continent joined North America.

Granite exposed in a roadcut one-half mile west of Sherman Pass. The bouldery outcrop is typical of granite.

Granite appears in roadcuts throughout the area on either side of Sherman Pass, gneiss in the area closer to Kettle Falls. Both kinds of rock consist mostly of pink and white feldspar, glassy gray quartz, and black mica or hornblende in various proportions. The mineral grains are generally large enough to distinguish with the unaided eye. As usual, the granite and gneiss resemble each other in consisting of the same minerals, but differ in that the gneiss is layered and streaky, and the granite is massive.

Flagstones await shipment at a quarry just north of Barney's Junction on Washington 20 west of Kettle Falls. The rock is a Paleozoic sandstone that splits into thin slabs because it as sheared in a fault zone along the east side of the Kettle dome.

The Kootenay Arc

The route between Kettle Falls and Tiger crosses the Kootenay arc. Rocks exposed along the road are tightly folded sedimentary formations, once part of the continental shelf, and a large mass of granite. As elsewhere in this part of Washington, the sedimentary rocks around the margins of granite intrusions contain ore bodies.

Remains of an old smelter beside Washington 20 about 5 miles east of Colville.

It is easy to recognize the sedimentary rocks because they are layered. Most of those along Washington 20 formed during Paleozoic time, between about 600 and 300 million years ago. A few fossils of animals that lived in the ocean during that time survived the folding well enough that geologists can still identify them, and so determine the ages of the rocks. The granite is easy to recognize too. It is massive gray rock that weathers into big, rounded boulders stained dark by iron oxide released as the rock weathers.

Crystal Falls, a picturesque waterfall and gorge eroded in granite beside Washington 20, about 15 miles east of Colville.

Strong parallel alignment of mineral grains makes the gneiss in these hills northeast of Omak platy, and that structure accounts for the flat, sloping faces on these hills. This area is on the western flank of the Okanogan dome.

Washington 155
Omak — Coulee Dam
54 mi. 87 km.

The road between Omak and the Grand Coulee Dam cuts across the southwestern corner of the Okanogan subcontinent. In the western 20 miles of the route, the road crosses old continental crust, rocks that once existed far out in the Pacific, and joined North America about 100 million years ago. This area is on the southwest side of the Okanogan dome where gneiss is the common rock in which the crystals are aligned parallel to each other, giving the rock a distinctive grain. Much of the gneiss also has dark and light colored bands streaking through it.

In the middle part of the route, as far east as Nespelem, the road crosses granite. It looks a bit like gneiss, and does contain the same minerals. But the mineral grains are not in parallel alignment, so the granite is not streaky, nor does it contain color bands. Watch for roadcuts in massive pale gray rock. This granite is less than 100 million years old, and formed after the Okanogan subcontinent had become part of North America. It is related to the trench that existed until about 50 million years ago where the Okanogan Valley now is.

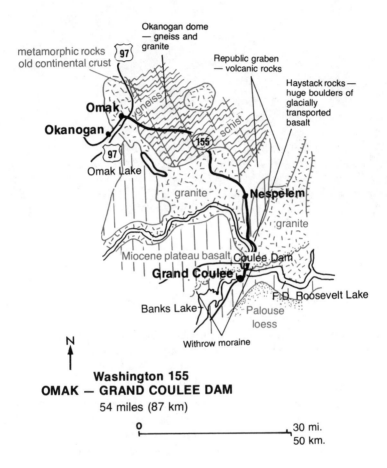

Omak

Okanogan

Omak Lake

metamorphic rocks
old continental crust

Okanogan dome
— gneiss and
granite

gneiss

schist

Republic graben
— volcanic rocks

Haystack rocks —
huge boulders of
glacially
transported
basalt

granite

Nespelem

granite

Miocene plateau basalt

Coulee Dam

Grand Coulee

Banks Lake

F.D. Roosevelt Lake

Palouse
loess

Withrow moraine

N

Washington 155
OMAK — GRAND COULEE DAM
54 miles (87 km)

0   30 mi.
            50 km.

A close view of tightly folded gneiss on the west flank of the Okanogan dome northeast of Omak.

Between the area just north of Nespelem and the Columbia River, the highway follows close to the west side of one of the dropped slices of crust that form the Republic graben. Rocks in the hills west of the road are granite like that exposed along the road west of Nespelem. Rocks in the hills east of the road are andesite and rhyolite that erupted about 50 million years ago.

Grand Coulee Dam stands where the basalt lava flows of the Columbia Plateau lap onto the older granite of the Okanogan subcontinent. Erosion of the Grand Coulee stripped the basalt right down to granite, thus exposing good foundation rock for the dam.

III
THE NORTH CASCADE
SUBCONTINENT

The North Cascades of Washington are the southern end of
the North Cascade subcontinent, which extends north through
much of British Columbia. Although most of the story is still
undeciphered, there is no doubt that the area was a small
continent that went its separate way somewhere in the vastness of the Pacific Ocean for many millions of years. The last of
the oceanic crust that formerly separated it from North
America vanished into the old Okanogan trench early in
Eocene time, about 50 million years ago.

Exotic Fossils

In a few places in Washington and British Columbia,
sedimentary rock formations more than 100 million years old
lie on top of the old continental crust. Those formations contain
the fossil remains of animals that differ from their relatives
farther east in about the same way that a modern collection of
sea shells from the coast of Washington would differ from a
collection gathered on the beaches of Australia. Unfortunately,
there are no places in Washington where people can see those
fossils near a major road.

The exposed part of the North Cascade subcontinent.

Geologists disagree about exactly where those exotic animals that left their remains in the North Cascades lived. But
their home must have been somewhere in the far reaches of the
tropical Pacific. Many of the North Cascade fossils resemble
their relatives of the same age in southeast Asia more closely
than those in North America east of the Cascades.

Two Oceanic Trenches

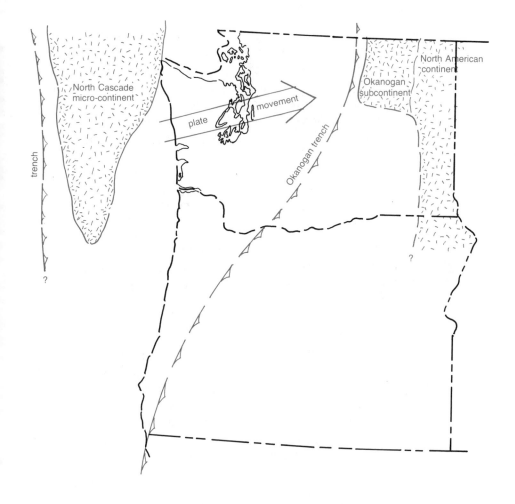

*A plate tectonic scenario to dock the North Cascade micro-continent
against western North America.*

Wherever the North Cascade micro-continent was a few hundred million years ago, it became involved with two oceanic trenches. One, the Okanogan trench, lay along the west coast of North America, passing through what is now the western part of the Okanogan Valley. The Okanogan trench had little direct effect on the Cascade micro-continent except to dictate its destiny as a future part of North America. Ocean floor sinking into the Okanogan trench drew the Cascade micro-continent to North America as inevitably as winding up the string lands a kite.

The other trench lay along the west coast of the micro-continent, moved along with it, and played a large part in creating the geologic scene we now see in the North Cascades. Most of the rocks now exposed in the western part of the range are old oceanic crust and its sedimentary cover that were stuffed into that trench. That western trench still swallows the floor of the Pacific Ocean off the modern west coast.

DOCKING THE NORTH CASCADE SUBCONTINENT

Inch by tedious inch, the old Okanogan trench along what was then the west coast of North America patiently consumed the expanse of ocean floor that separated the North Cascade micro-continent from North America. Generations of dinosaurs living in the Okanogan highlands, which were already part of North America, must have seen the island continent gradually rise on the western horizon, approaching almost imperceptibly closer with each succeeding generation. Most geologists think the approaching micro-continent did not approach North America directly, but glided slowly northeastward, like a great ship standing obliquely into port. One reason for thinking that is the generally tropical look of the fossils in the North Cascade subcontinent.

Then, about the time the dinosaurs vanished, the projecting headlands of North America and the micro-continent slowly met, like outstretched fingers, as the ocean separating them dwindled to a narrow gulf. As that happened, the layers of sedimentary rocks that formed the coastal plains on both sides of the micro-continent crumpled, like a squeezed accordion, into folds that trend generally northwest — another reason for believing that the island approached from the southwest. Finally, the last intervening scrap of ocean floor vanished into the trench, and the wandering island docked against the west coast of North America to become the North Cascade subcontinent.

The Okanogan trench destroyed itself as it swallowed the last of the heavy oceanic rocks, and brought the North Cascade micro-continent onto North America. That happened simply because trenches exist only because a plate is sinking, and continental crust is too light to sink. However, the trench that had existed for a long time on the west coast of the subcontinent continued to swallow the floor of the Pacific Ocean.

The Old Coastal Plain

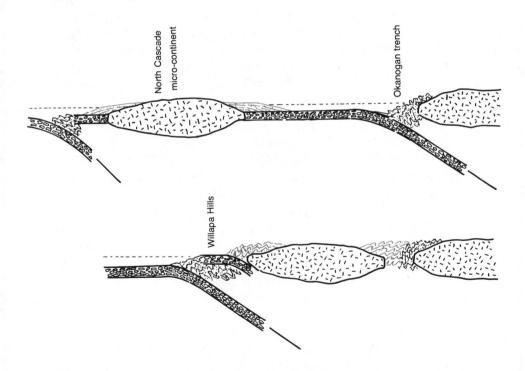

The collision between the North Cascade micro-continent and North America jammed the coastal plains into tight folds.

Sedimentary rocks on both the east and west sides of the North Cascade subcontinent accumulated along its coast while it was still an island in the Pacific. Most of the rocks contain volcanic debris, so we have solid evidence of an island microcontinent complete with erupting volcanoes. Those dinosaurs that saw the island continent approach probably watched volcanic eruptions on the distant western horizon as the North Cascade subcontinent slowly rose over the horizon, at a rate of 2 or 3 inches per year.

Some of the sedimentary rocks around the old coast formed in shallow water offshore, and contain abundant faunas of fossil sea shells. Other formations consist of sediment laid down on land, and those contain plant fossils and coal. Sedimentary formations near Seattle, on the west coast of the North Cascade subcontinent, have produced coal since the last century, as have other formations of the same age near Wenatchee and Cle Elum, on the former east coast. Like most large coal seams, these appear to have formed in tidewater swamps. So the coastal plains of the North Cascade micro-continent contained vast swamps and marshes, in which partially decayed vegetation accumulated to form deposits of peat. Picture a large island, with a chain of active volcanoes along its crest and lush forests along its coasts.

Thick deposits of sandstone and shale, which geologists call the Swauk formation, blanket large areas along the eastern margin of the North Cascade subcontinent. The formation is thousands of feet thick, and age dates show that it all accumulated between about 65 and 50 million years ago, while the micro-continent was closing in to North America. The age of the Swauk sandstone, and its position along the eastern margin of the Cascade subcontinent, suggest that it accumulated in the last narrowing gap between the approaching North Cascade micro-continent and North America.

Folds

All the old coastal plain sedimentary formations are now tightly folded, evidently because the collision between the North Cascade micro-continent and North America crumpled them. Most of the folds trend northwest throughout large areas on both the west and east sides of the subcontinent. That is the pattern you would expect if the micro-continent came in from the southwest. If you kick a rug, it wrinkles in a direction transverse to that of the kick.

Most of that folding happened about the middle of Eocene time. The tightly crumpled sedimentary rocks were deposited as flat lying layers early in Eocene time. Late Eocene volcanic rocks are much less folded than the sedimentary rocks of the old coastal plain, so most of the folding had happened before they erupted. Evidently, the micro-continent was still offshore early in Eocene time, crushed its coastal plain as it jammed in to North America during mid-Eocene time, and had docked by late Eocene time.

The Straight Creek Fault

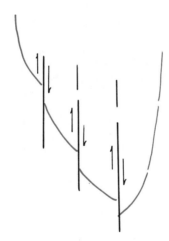

Northward wrenching of the landing North Cascade micro-continent broke it into slices.

The North Cascade micro-continent approached North America obliquely from the southwest because the plate in which it was embedded was moving that way. That northward urging continued after the former eastern coast of the micro-continent was firmly anchored to the west coast of North America along the line of the former oceanic trench, just west of the Okanogan Valley.

The grounded subcontinent responded to the northward movement of the plate in which it had been riding by breaking along faults that trend from north to south, and from northwest to southeast. The north to south trending faults split the docked subcontinent into long slices, each of which slid north relative to the next slice to the east. The Straight Creek fault is the longest and most prominent of those fractures. Some geologists estimate that the continental slice west of the Straight Creek fault slid as much as 150 to 200 miles north of the slice to the east, and most agree that it moved at least 100 miles.

The Methow and Chiwaukum Grabens

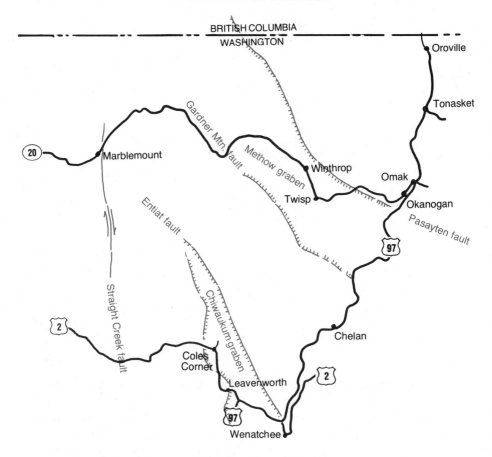

The Chiwaukum and Methow grabens

Northward wrenching of the North Cascade subcontinent along the faults that trend from northwest to southeast opened the Chiwaukum graben in the area generally northwest of Wenatchee, and the Methow graben along the Canadian border. Both are large blocks that moved between parallel faults that trend northwest. Although both segments of the earth's crust dropped relative to the adjacent areas, the faults that bound them also appear to have moved horizontally. The Methow and Chiwaukum grabens look on the geologic map like long pie slices slivered into the eastern part of the North Cascade subcontinent.

Geologists call the rocks that fill the Chiwaukum graben the Chumstick formation. A number of convincing age dates show that most of the Chumstick formation accumulated between 40 and 50 million years ago. If the big faults in the North Cascades moved as the North Cascade subcontinent landed against North America, then the age of the Chumstick formation dates that event. The evidence of the Chumstick formation, like that of the folded coastal plain formations, shows that the micro-continent was landing between 40 and 50 million years ago, during Eocene time.

Rocks in the Methow graben include Jurassic and Cretaceous sedimentary formations, some of which contain large numbers of fossil sea shells. Those rocks, which are now tightly folded, certainly accumulated on a shallow continental shelf before the North Cascade micro-continent joined North America. Some geologists believe that the minerals in the sandstones came from the Okanogan highlands. If so, then the sedimentary rocks in the Methow graben accumulated about where we now see them. On the other hand, their position in the eastern margin of the North Cascade subcontinent suggests that the sedimentary rocks in the Methow graben may have accumulated along its coast, while the micro-continent was still far out in the Pacific. Further study of the fossils may resolve the issue by revealing how closely they resemble North American fossils of the same age.

THE OLD CONTINENTAL CRUST

Large parts of the North Cascade subcontinent east of the

Straight Creek fault consist of old continental crust. Rocks in those areas are complex assemblages of granites, gneisses and schists, the kinds of rocks that form the floating rafts we call continents.

It is extremely difficult to estimate the ages of the old continental crustal rocks in the North Cascade subcontinent. None contain fossils that might provide a clue, and age dates yield conflicting results. The basic problem is that these rocks were all reheated enough to reset, or partially reset, their radioactive clocks. So the age dates tell a mixed story compounded in part of the actual age of the rock, in part of the time since it was last reheated. However, there is no doubt that much of the old continental rock in the North Cascade subcontinent is more than 100 million years old, and good reason to suspect that some may be much older.

THE WESTERN TRENCH AND THE YOUNGER ROCKS

While the Okanogan trench patiently devoured the ocean floor that separated the North Cascade subcontinent from North America, the Cascade trench was also swallowing sea floor off the west coast of the micro-continent. It created many of the rocks we see in the North Cascades today long before the region joined North America.

The Trench Filling

Approximately the western third of the North Cascade subcontinent, the part west of the Straight Creek fault, consists mostly of rock stuffed into the trench that existed along its western coast. Like all trench fillings, these rocks are basically oceanic in origin. Most are sedimentary rocks that formed on the ocean floor; some are basalt of the oceanic crust.

In general, rocks west of the Straight Creek fault become younger and less metamorphosed westward, no doubt because the trench filled from east to west. The oldest rocks closest to the Straight Creek fault include sedimentary formations that accumulated on the ocean floor more than 200 million years ago, as well as generous slices of the ocean floor itself. Those rocks have been extremely hot, and are now so thoroughly metamorphosed that they hardly resemble the muddy sandstones and pillow basalts that were stuffed into the trench. Radioactive ages of the minerals formed during metamorphism give dates of more than 100 million years, so we must conclude that the rocks were deep in the trench then. And that was about 50 million years before the micro-continent landed.

Farther west, the trench filling consists of younger and less recrystallized rocks, material that still looks like sedimentary rock. However, their severely deformed condition, and some of the minerals they contain, leave no doubt that the younger rocks were also stuffed into a trench. Some of those western rocks are blueschists, so called because they contain several blue minerals.

Blueschists are rare, and were for many years one of the most perplexing problems in geology. Experimental work showed that the blue minerals form only under conditions of extremely high pressure, and moderately high temperature. No one could imagine how a rock could get deep enough below the surface for blueschist minerals to form without also getting so hot that they could not form. It is now clear that blueschist minerals form in rocks stuffed so rapidly into a trench that they get deep enough for the characteristic blue minerals to form long before they get hot enough to prevent their forming. If those rocks return to shallow depth before they get hot, the blue minerals survive, and we find blueschists in the roadcuts.

Granite and Related Rocks

Wherever a slab of lithosphere sinks beneath continental crust, molten basalt and steam rising from the sinking slab melt the hot rocks in the lower part of the continental crust, converting them into molten granitic magma. In some cases, the trench filling itself gets hot enough to melt into granitic magma. Remember that sediments have the same general chemical composition as continental crustal rocks, so they turn into granitic magma when they melt.

Most of the masses of granitic magma that melted in the lower part of the North Cascade subcontinent crystallized below the surface to form plutons of granite or closely related rock types. Relatively few erupted through volcanoes to form rhyolite or andesite. The North Cascade subcontinent is full of granite plutons. Age dates show that most of them formed during the last 200 million years, so they must be the progeny of the trench that has existed off the west coast of the subcontinent for the last 200 million years. In a general way, the oldest granites tend to lie in the eastern part of the subcontinent, and their ages become younger westward to a minimum of about 15 million years.

Some of the younger granite plutons cut right across the Straight Creek fault, as well as the faults that bound the Methow and Chiwaukum grabens. The faults do not offset those younger granites, so they can not have moved since the granite plutons invaded the rocks on either side of them. That relationship adds weight to the argument that the Straight Creek fault and its cohorts moved during the time the North Cascade subcontinent was joining North America, and stopped moving when docking was complete.

Volcanoes

Volcanoes were active on the North Cascade micro-continent long before it joined North America, and have continued to erupt there most of the time since. It seems reasonable to argue that volcanic rocks more than approximately 50 million years old erupted while the North Cascade micro-continent was still a large island off the west coast of North America. They belong to a volcanic chain that existed somewhere offshore. The younger volcanic rocks that erupted after the North Cascade subcontinent docked are part of the Cascade chain, and have always belonged to North America.

U.S. 2
Everett — Wenatchee
128 mi. 205 km.

The North Cascade Subcontinent

Between Everett and Wenatchee, U.S. 2 crosses the North Cascade subcontinent from its former western shore at the edge of the Puget Sound lowland to its old eastern shore near Wenatchee. Some of the rocks along the way formed while the subcontinent was still an island in the Pacific Ocean; others formed after it sutured on to North America about 50 million years ago. Except near Wenatchee, the landscapes visible from this road owe their origin more to mountain glaciation than to any other process.

About 5 miles east of Skykomish, approximately halfway between Everett and Wenatchee, the highway crosses the Straight Creek fault. The fault trends almost directly north, and geologists follow it from the area west of Yakima north through much of British Columbia. It divides the North Cascade subcontinent into two distinct provinces.

In general, rocks west of the Straight Creek fault are oceanic. They consist of oceanic crust and of mudstones and muddy sandstones that accumulated on the ocean floor. All were stuffed into the trench along the west side of the micro-continent long before it joined North America. Now they are all severely deformed, and considerably recrystallized into metamorphic rocks. Rocks east of the Straight Creek fault consist partly of old continental crust, partly of sedimentary rocks deposited along the eastern coastal plain of the North Cascade subcontinent.

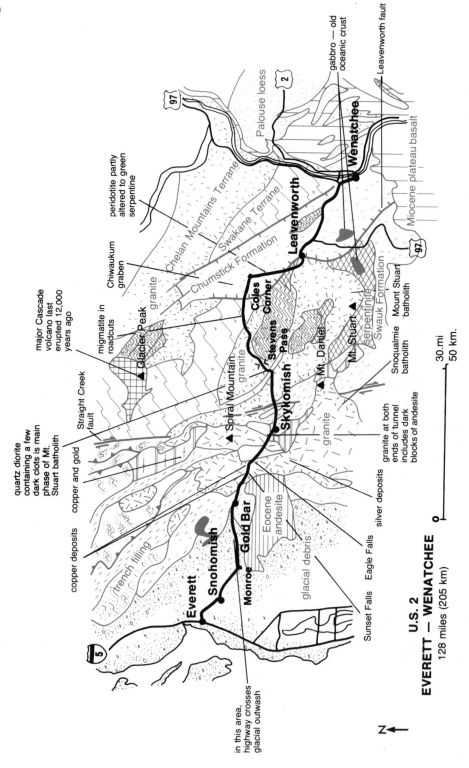

gabbro — old
oceanic crust

Leavenworth fault

peridotite partly
altered to green
serpentine

Palouse loess

Wenatchee

Chelan Mountains Terrane

Swakane Terrane

Leavenworth

Miocene plateau basalt

Chiwaukum
graben

Chumstick Formation

major Cascade
volcano last
erupted 12,000
years ago

migmatite in
roadcuts

granite

Glacier Peak

Coles
Corner

Stevens
Pass

Mt. Stuart

serpentinite

Swauk Formation

Mount Stuart
batholith

Snoqualmie
batholith

Straight Creek
fault

Spiral Mountain

granite

Skykomish

Mt. Daniel

granite

granite at both
ends of tunnel
includes dark
blocks of andesite

silver deposits

quartz diorite
containing a few
dark clots is main
phase of Mt.
Stuart batholith

copper and gold

copper deposits

trench filling

Everett

Snohomish

Gold Bar

Monroe

Eocene
andesite

glacial debris

Sunset Falls   Eagle Falls

in this area,
highway crosses
glacial outwash

U.S. 2
EVERETT — WENATCHEE
128 miles (205 km)

0   30 mi
         50 km.

N ←

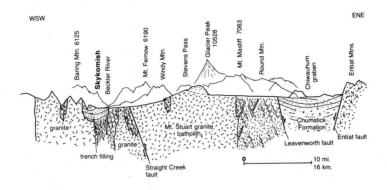

Section across the crest of the North Cascade subcontinent approximately along the line of U.S. 2.

ROCKS WEST OF THE STRAIGHT CREEK FAULT

U.S. 2 follows the Skykomish River all the way through the area west of the Straight Creek fault. In general, the rocks are youngest and least metamorphosed near Everett, and become progressively older and more metamorphosed eastward. Therefore, geologists have found it convenient to divide them into western and eastern zones.

The Western Metamorphic Zone

Western zone rocks appear along the road from the western margin of the North Cascade subcontinent near Monroe to the general vicinity of Gold Bar. Most of those rocks are gray sandstones and muddy sandstones, but the zone also includes a small area of broken up oceanic crust, so there is little doubt that all the rocks in it are oceanic. All the rocks are folded and broken, but only very slightly metamorphosed. Evidently, they were not buried deeply in the trench long enough to get very hot.

In fact, some of the western zone rocks are so slightly altered that they still contain the fossil remains of animals that lived during late Jurassic and early Cretaceous time, approximately 150 million years ago. That must have been when these sandstones accumulated on the ocean floor.

The Eastern Metamorphic Zone

East of the Gold Bar area, the rocks become somewhat more

thoroughly metamorphosed, much more thoroughly deformed, and pass into the eastern metamorphic zone. It contains large quantities of black metamorphic rocks, old basaltic oceanic crust, and relatively little of the muddy sandstone so abundant in the western zone. The eastern zone rocks also contain some marble, metamorphosed limestone.

Most of the rocks in the eastern metamorphic zone are so thoroughly deformed and recrystallized that any fossils they once contained are gone. Nevertheless, a few fossils do survive here and there, and they include the remains of animals that lived as long ago as Permian time, more than 200 million years ago. Clearly, these rocks formed on the ocean floor long before those in the western metamorphic zone. Radioactive dates on minerals formed during metamorphism give Cretaceous ages, about 100 million years. So these rocks were stuffed into the trench at about that time.

Granite Intrusions

A close look at granite. The black grains are flakes of biotite mica and needles of hornblende. The white mineral is feldspar, and the gray grains are quartz. All the mineral grains are about the size of split peas.

Granite intrusions exist in both the eastern and western metamorphic zones. Age dates on some masses of granite intruded into the western zone rocks give ages of about 45 million years, Eocene time. That was when the North Cascade subcontinent was annexing itself to North America. Those western zone granites show no sign of deformation, so it seems that the surrounding rocks had already been stuffed into the trench when masses of molten granite magma intruded them.

The eastern metamorphic zone contains numerous granite intrusions, most of them apparently emplaced after the North Cascade subcontinent annexed itself to North America. All are pale gray rocks speckled with crystals of black biotite. Those the road crosses are typical of others back in the hills.

For a distance of about 5 miles in the area west of Baring, U.S. 2 passes through the southern end of the Index granite batholith. Age dates show that it is about 33 million years old, an Oligocene age. The volcanoes of the Western Cascades were active then. Farther east, the road crosses the southern tip of the Grotto granite batholith for a distance of about a quarter of a mile on either side of the tunnel west of Skykomish. Roadcuts near the tunnel are full of dark inclusions of older volcanic rock that the granite magma picked up as it rose. The Grotto granite is 24 million years old, a Miocene age that also relates to the Western Cascades. Some of the younger volcanic rocks are rhyolite ash with about the same composition as the Grotto batholith, and may well have erupted from it. The Beckler Peak batholith, another intrusion related to the Western Cascades, shows up in roadcuts in the area about 3 miles east of Skykomish, just west of the Straight Creek fault.

The closest modern volcano, Glacier Peak, is about 25 miles directly north of Stevens Pass. Its last eruption, about 12,000 years ago, spread rhyolite ash over much of the Pacific Northwest. It seems safe to assume that some millions of years from now erosion will reveal a granite batholith similar to the Index and Grotto batholiths where Glacier Peak now stands.

ROCKS EAST OF THE STRAIGHT CREEK FAULT

East of the Straight Creek fault, generally from the area about 13 miles west of Stevens Pass to Leavenworth, the road crosses a complex assortment of thoroughly metamorphosed sedimentary rocks, which contain large masses of intrusive granitic rocks. The metamorphic rocks are probably older continental crust, perhaps hundreds of million of years old. The masses of granite are much younger than the rocks that contain them, but were already there long before the Cascade micro-continent joined North America.

The first step geologists take in starting to cope with such messy complexes of metamorphic and igneous rocks is to divide them into areas called terrains, in which the various rocks appear to have something in common. Just east of the Straight Creek fault, the road enters the Nason Ingalls terrain, which it crosses all the way to Leavenworth. The Nason Ingalls terrain consists mostly of metamorphic rocks, gneiss and schist, which appear along the road west of the pass in the valley of Nason Creek, between Merritt and Coles Corner. The gneiss is very streaky, full of dark and light bands, which were originally sedimentary layers.

The Mount Stuart Batholith

U.S. 2 crosses part of the Mount Stuart batholith in the area on both sides of Stevens Pass, another part just west of Leavenworth about 35 miles farther east. Watch for roadcuts in gray granite. Most of the rock is massive, homogeneous, and uniformly gray, but it becomes streaky and full of dark inclusions near the margins. The two masses of nearly identical granite are very close together, and probably merge at depth into a continuous body of rock. Age dates show that this granite is about 88 million years old, and so formed during Cretaceous time.

The Chiwaukum Graben

Between Leavenworth and Wenatchee, the highway crosses the Chiwaukum graben, a large block of the earth's crust that dropped along faults during Eocene time. On the geologic map, the Chiwaukum graben looks as though a long, northwest trending slice about 12 miles wide was chopped out of the older rocks of the North Cascades, then filled with younger sedimentary rocks, the Chumstick formation. Actually, the older rocks are still there, hidden beneath the younger sedimentary rocks.

The Leavenworth fault zone, which the road crosses at Leavenworth, and follows between Leavenworth and Coles Corner, bounds the south side of the Chiwaukum graben. The Chumstick formation shows up here and there along the road between Leavenworth and Wenatchee, and forms the hills west of Wenatchee. It is very pale rock, which consists largely of volcanic debris.

U.S. 97
Cle Elum — U.S. 2
42 mi. 67 km.

U.S. 97 follows the railroad several miles southeast from Cle Elum, and then turns northeast to follow the lower Teanaway River for several miles to Swauk Prairie, a lovely rolling area underlain by several feet of wind-blown silt deposited on top of glacial till that dates from one of the earlier ice ages.

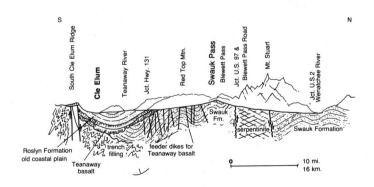

Section along the line of U.S. 97 between Cle Elum and the junction with U.S. 2.

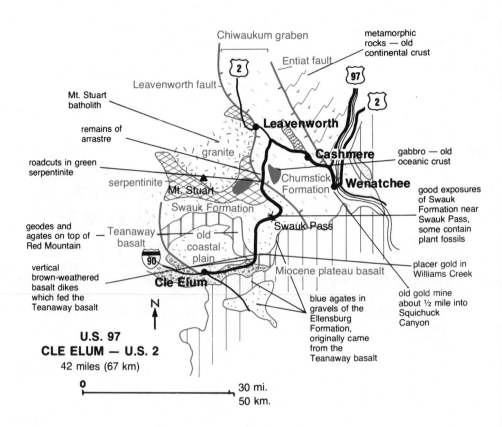

Chiwaukum graben

metamorphic
rocks — old
continental crust

Entiat fault

Leavenworth fault

Mt. Stuart
batholith

remains of
arrastre

roadcuts in green
serpentinite

granite

serpentinite

Mt. Stuart

Swauk Formation

geodes and
agates on top of
Red Mountain

Teanaway
basalt

old
coastal
plain

vertical
brown-weathered
basalt dikes
which fed the
Teanaway basalt

Cle Elum

Leavenworth

Cashmere

Chumstick
Formation

Wenatchee

gabbro — old
oceanic crust

good exposures
of Swauk
Formation near
Swauk Pass,
some contain
plant fossils

Swauk Pass

Miocene plateau basalt

placer gold in
Williams Creek

old gold mine
about ½ mile into
Squichuck
Canyon

blue agates in
gravels of the
Ellensburg
Formation,
originally came
from the
Teanaway basalt

N

U.S. 97
CLE ELUM — U.S. 2
42 miles (67 km)

0   30 mi.
                     50 km.

Plateau Basalt

Lookout Mountain, the hill south of the highway where it follows the Teanaway River and crosses Swauk Prairie, has a flat top because it is capped with a lava flow, one of the Columbia Plateau flood basalts. That lava flow winds through this area in a snaking pattern, evidently because it filled an old winding stream valley. Now that former valley is a ridge because the basalt resists erosion more successfuly than the rocks around it. Geologists call such situations "inverted topography."

The Swauk Formation

At the north end of Swauk Prairie, the road follows Swauk Creek through a valley eroded in the Teanaway basalt, which erupted during Eocene time, about 50 million years ago. The road continues up Swauk Creek through a thick sequence of sandstones of the Swauk formation, which continues most of the way to the junction with U.S. 2.

Tilted layers of light gray sandstone and dark gray shale of the Swauk formation exposed beside U.S. 97 just south of Swauk Pass.

The Swauk formation contains at least several thousand feet of sandstone, pebble conglomerate, shale and assorted other sedimentary rock types. Most of the outcrops along the road are brownish sandstone. At least part of the formation appears to have accumulated on land, because the patterns of internal layering in some of the sandstones are like those that typically form in streams. All of the Swauk formation was deposited along the eastern margin of the North Cascade micro-continent during Eocene time, probably a bit more than 50 million years ago. It is about the same age as the Eocene coastal plain deposits around the margins of the subcontinent, and must have accumulated just as the micro-continent was about to dock.

As the North Cascade subcontinent jammed into North America obliquely from the southwest, the collision crumpled the Swauk formation into folds, which trend generally northwest. The lava flows of later Eocene Teanaway basalt near Cle Elum are much less folded, so the collision must have happened sometime between deposition of the Swauk formation and eruption of the Teanaway volcanic rocks. Remember that Eocene time includes some 10 million or so years, time enough for quite a few events.

The Mount Stuart Batholith

The spectacular jagged mountans in the distance west of the road north of Swauk Pass are the Stuart Range, the glacially eroded Mount Stuart granite batholith. Age dates show that the granite is about 88 million years old, and must therefore have formed during late Cretaceous time, before the North Cascade subcontinent joined North America.

The Ingalls Complex

Rocks around the southern part of the Mount Stuart batholith consist of a severely deformed slab of old oceanic crust that geologists include in the Ingalls metamorphic complex. It appears to have been stuffed into a trench before granite magma invaded the area to form the batholith. The road crosses the eastern part of that complex for several miles in the area from the old town of Blewett, about 10 miles north of Swauk Pass, north to the mouth of Ingalls Creek.

As usual, most of the oceanic crustal rocks are dark green and black. In the Blewett area, they include large masses of an unusual rock called serpentinite, which forms when the peridotite of the mantle reacts with water. Some people call serpentinite "soapstone"

*A roadcut in dark green serpentinite near Blewett, about 10 miles
north of Swauk Pass. The greasy looking rock consists of a mass of
chunks with slick surfaces polished as they slipped past each other
during deformation.*

because it has a slippery feel. Most serpentinite is soft enough to
carve with an ordinary pocketknife, and it often shows up in souvenir
shops as little figurines and boxes.

Serpentinite is as slippery in the earth's crust as its soapy feel
might make you suspect. Some serpentinite masses appear to have
gotten where they are by intruding the rocks that enclose them. The
slippery serpentinite simply squeezes through any weak places it can
find in the earth's crust as though it were blobs of grease. Many
geologists refer to that kind of movement as "watermelon seed
tectonics."

Gold

*Gold dredge tailings
beside U.S. 97 a few
miles south of its
junction with U.S. 2.*

Prospectors found placer gold in lower Swauk Creek in 1874, lode deposits in bedrock a few years later. Those discoveries started a mining boom that still reverberates whenever the price of gold goes up.

Early work with manual labor depleted the easy placer pickings, and then two large dredges worked the stream gravels during the 1920's, leaving the dredge spoil heaps that still stand beside the road. However, the dredges made more of a mess than a profit, and quit after having worked only a short section of the creek bottom. Nevertheless, the district is known for having produced a few real bonanzas, rich pockets of uncommonly large nuggets, and to this day people occasionally still find such windfalls. This is one of the few areas of Washington where weekend prospectors can actually hope to find more than healthy outdoor exercise.

Blewett was quite a gold mining center during the decades just before and after the turn of the century. Most of the gold came from placer deposits in Peshastin Creek. However, several underground mines worked gold quartz veins, and the town had a large stamp mill to crush the ore and wash the gold free. The remains of a much older mill, an arrastre, still survive at a marked site beside the road. Many early gold mining camps had arrastres, which used water or horse

An old arrastre about 10 miles south of Swauk Pass.

power to drag large stones around on a circular stone pavement to grind the ore. The method looks primitive, but arrastres were cheap, and they worked.

Ingalls Creek, which enters Peshastin Creek about 4 miles north of Blewett, is the setting for a lost mine story in the best tradition of such accounts. It seems that a cavalry officer named Ingalls found a rich vein of gold-bearing quartz beside one of three distinctive small lakes in the headwaters of the creek, high in the glaciated Mount Stuart Range. While he was returning from that trip, an earthquake shook the region triggering many landslides. Indians killed Captain Ingalls before he had chance to return, and a friend, to whom he had carefully described the place, continued the search. However, neither the friend nor anyone else has ever found those three lakes. Many people believe that the landslides may have changed the setting beyond recognition. Many others think it is simply a good story.

The Chiwaukum Graben

Just north of Ingalls Creek, the road crosses the Leavenworth fault zone, and enters the Chiwaukum graben, which it follows to Wenatchee. It is a slice of the earth's crust that slipped down between the Leavenworth fault and the parallel Entiat fault, which passes through the north edge of Wenatchee.

Rocks within the Chiwaukum graben consist largely of the Chumstick formation, which the road crosses along most of the route between the mouth of Ingall's Creek and U.S. 2. The hills directly west of Wenatchee also consist mostly of the Chumstick formation. Most of the Chumstick formation consists of sandstone so pale it is almost white, and it also includes large quantities of volcanic material.

The Leavenworth and Entiat faults both cut the Swauk formation, so it must be older than the Chiwaukum graben. Radioactive age dates on the Chumstick formation within the Chiwaukum graben show that it accumulated between about 35 and 40 million years ago, late Eocene time. The Chumstick formation is folded, but much less tightly so than the older Swauk formaton. Therefore, it appears that most of the deformation occurred before the Chiwaukum graben filled with volcanically derived sedimentary rocks during late Eocene time. If the North Cascade subcontinent did indeed cause the folding as it jammed into North America, then it must have docked during the middle and latter part of Eocene time.

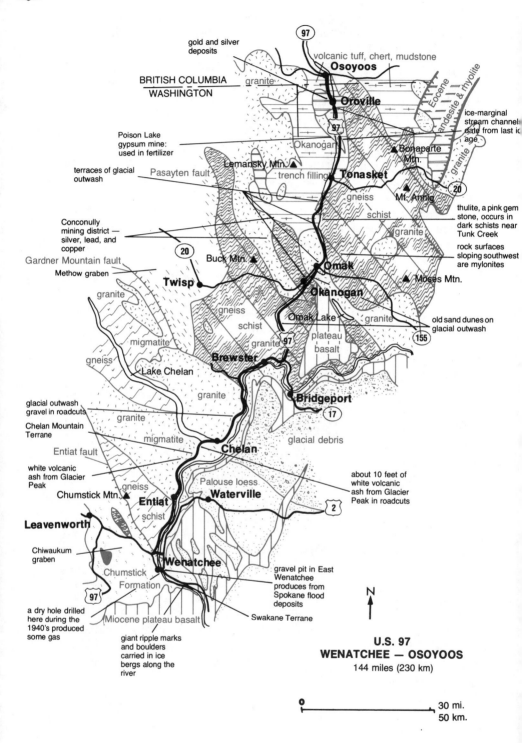

gold and silver deposits

volcanic tuff, chert, mudstone

Osoyoos

granite

Oroville

BRITISH COLUMBIA
WASHINGTON

Eocene

andesite & rhyolite

ice-marginal stream channel date from last ice age

Poison Lake gypsum mine: used in fertilizer

Okanogan

Bonaparte Mtn.

granite

terraces of glacial outwash

Pasayten fault

Lemansky Mtn.

trench filling

Tonasket

Mt. Annig

gneiss

schist

thulite, a pink gem stone, occurs in dark schists near Tunk Creek

granite

Conconully mining district — silver, lead, and copper

rock surfaces sloping southwest are mylonites

Gardner Mountain fault

20

Buck Mtn.

Omak

Moses Mtn.

Methow graben

Twisp

Okanogan

granite

gneiss

schist

Omak Lake

granite

old sand dunes on glacial outwash

155

migmatite

granite

97

plateau basalt

gneiss

Brewster

Lake Chelan

granite

Bridgeport

glacial outwash gravel in roadcuts

granite

17

Chelan Mountain Terrane

migmatite

Chelan

glacial debris

Entiat fault

Palouse loess

white volcanic ash from Glacier Peak

gneiss

Waterville

about 10 feet of white volcanic ash from Glacier Peak in roadcuts

Chumstick Mtn.

Entiat

schist

2

Leavenworth

Chiwaukum graben

Wenatchee

Chumstick Formation

gravel pit in East Wenatchee produces from Spokane flood deposits

97

a dry hole drilled here during the 1940's produced some gas

Miocene plateau basalt

Swakane Terrane

N

giant ripple marks and boulders carried in ice bergs along the river

U.S. 97
WENATCHEE — OSOYOOS
144 miles (230 km)

0   30 mi.
50 km.

82

U.S. 97
Wenatchee — Osoyoos, B.C.
144 mi. 230 km.

All the way between Wenatchee and Osoyoos, the highway skirts the eastern margin of the North Cascade subcontinent. The route follows the Columbia River between Wenatchee and Brewster, along the boundary between the North Cascade subcontinent and the Columbia Plateau. Between Brewster and Osoyoos, the road follows the Okanogan trench, with the highlands of the North Cascade subcontinent rising in the west, those of the Okanogan subcontinent in the east. All of the country north of Chelan was buried under the Okanogan lobe of the big ice sheet during the last ice age, and the Spokane floods scoured the Columbia River valley south of Chelan.

Spokane Flood Deposits

Most of Wenatchee stands on alluvial fans composed of gravel swept out of the valleys west of town. The higher parts of East Wenatchee and Pangborn Air Field are on a much younger surface mantled with gravel deposits left by the Spokane flood. Look carefully to see giant current ripples as much as 30 feet high, and several hundred feet from crest to crest that gently undulate, like very long waves. A lower surface near the river also has giant current ripples on it. They must have formed during later Spokane floods, which seem to have been smaller than the earlier versions. Two or three feet of windblown dust accumulated since the last of the Spokane floods cover the giant current ripples on the higher terrace.

The Columbia River

Between Wenatchee and Brewster, the highway follows the west bank of the Columbia River as it traces the boundary between the Columbia Plateau to the east and the North Cascade subcontinent to the west. Evidently, the river began flowing along the low outermost edge of the volcanic plateau, where the lava flows lapped onto the edge of the highlands.

The Columbia River has now cut down through the plateau basalt into the older igneous and metamorphic rocks of the North Cascade subcontinent, which are exposed along the road. Dark bluffs in the distance across the river are plateau basalt. Rocks exposed along the highway on the other side of the river are essentially the same as those along U.S. 97.

The Roaring Creek Terrain

Dikes of white pegmatite, a coarse-grained variety of granite, fill fractures in dark granodiorite, a rock closely related to granite. This exposure is in a roadcut just north of Wenatchee, part of the Roaring Creek terrain.

Between the Entiat fault just north of Wenatchee and the town of Entiat, a distance of about 16 miles, the road passes through a complex of igneous and metamorphic rocks, which some geologists call the Roaring Creek terrain. The word terrain is one of those quaint geologic expressions which simply means an area in which the rocks appear to have a lot in common.

Most of the Roaring Creek terrain rocks exposed along the road are the Swakane gneiss. It is fairly easy to recognize as pale gray and streaky looking rock composed mostly of milky feldspar, shadowy grains of glassy quartz, and flakes of black biotite mica. Age dates on the Swakane gneiss give confusing and contradictory numbers, which geologists have interpreted in several ways. However, it seems most likely that the rock metamorphosed sometime around 100 million years ago. There is also some dispute about whether the Swakane gneiss was volcanic or sedimentary rock before it was deeply buried and recrystallized at a red heat to convert it into the gneiss we see today.

The Swakane gneiss appears to grade upward into a mass of very fancy looking metamorphic rocks full of tightly folded light and dark layers. The layering leaves no doubt that this was sedimentary rock before metamorphism converted it into glittering schists full of flaky grains of mica, dark gneisses called amphibolite because they consist largely of black hornblende amphibole, and pale layers of streaky looking gneiss.

Chelan Mountains Terrain

Geologists lump the varied bedrock between Entiat and the area north of Chelan into the Chelan Mountains terrain. It consists mostly of igneous rock crystallized from magmas. Although the igneous rocks of the Chelan Mountains terrain superficially resemble the Swakane gneiss, it is easy to tell them apart. Igneous rocks are generally homogeneous, and they have a massive texture quite unlike the streaky grain typical of metamorphic rocks.

Most of the igneous rock in the Chelan Mountains terrain arc is pale gray granite composed mostly of milky feldspar, along with glassy grains of quartz, and black crystals of needle shaped hornblende or flaky biotite mica. Radioactive age dates show that the granite crystallized sometime between 60 and 70 million years ago.

Near Chelan, the rocks become migmatite, a streaky mixture of igneous and metamorphic rock swirled together to give a marble cake effect. Many migmatites form as metamorphosing rocks get hot enough to begin to melt. Granite magma melts at the lowest

Dikes of white pegmatite fill fractures in swirling dark and light migmatite in a roadcut near Chelan.

temperature, so the igneous part of such migmatites is always gray or pink granite. The dark swirls are shreds and fragments of metamorphic rock that remained solid as the moving granite magma swept it along.

The migmatite around Chelan appears to grade into the more uniform and completely igneous granite farther south. That leads geologists to suspect that the granite magma farther south may have formed about where we now see it, as a mass of metamorphosing rock got hot enough to melt.

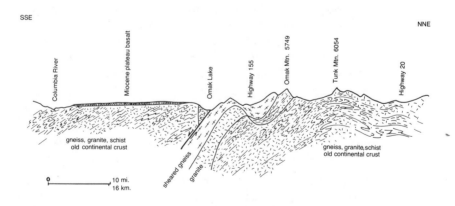

Section along a line east of U.S. 97 between Brewster and Tonasket.

Lake Chelan

The town of Chelan stands on a big moraine deposited around the lower end of an enormous glacier that filled the valley during the last ice age. The moraine functions as a natural earth fill dam to impound Lake Chelan. Views up the lake to the distant mountains are fantastic, primarily because the glacier gouged the valley almost straight. Ice does not go around bends as easily as water, so large glaciers straighten their valleys as they gouge them out.

Glaciation

The big glaciers of the last ice age covered all the northern fringe of Washington, and advanced down the western part of the Okanogan Valley about as far south as Chelan to produce some of the most spectacular glacial topography in Washington. The Withrow moraine that stretches in a broad tract of lumpy hills from Chelan southeast across the Columbia Plateau to Coulee City records the farthest reach of the ice. Only the highest hills in the southern part of the glaciated area stood like islands above the groaning sea of slowly moving glacial ice.

The prominent flat-topped bench in the middleground is a terrace composed of glacial outwash deposited as the ice melted between 10 and 12 thousand years ago.

All evidence everywhere tells of the last ice age ending with an abrupt change in climate. The final retreat of the glaciers was not so much an orderly withdrawal as an unseemly rout. Indians who almost certainly lived in the Okanogan Valley during those years must have marveled at the annual torrents of muddy meltwater, at the shocking northward shrinkage of the ice, and at the new expanses of muddy sediment left behind each year. It was a time of rapid environmental change, which must have seemed catastrophic to the people who watched it.

Except for the apple trees, the scene remains much as the melting ice and the floods of meltwater left it when the last ice age ended about 10 or 12 thousand years ago. The smooth surfaces of broad fans of sediment still remain, as do deep channels that once carried torrents of muddy water on warm summer afternoons when the ice was melting. The glacial stream channels are dry now, as are the channels of streams that flowed this way before the ice ages, and no more floods of muddy meltwater pour across the sediment surfaces. But the basic elements of the ice age landscape remain for anyone to see. You just need to squint hard enough to blank out the trees and the buildings.

An abandoned glacial meltwater channel beside the highway north of Riverside. A summer river of meltwater roared through this valley as the ice melted back at the end of the last ice age. It will remain dry until the next ice age.

The Okanogan glacier eroded these sloping steps in the bedrock beside U.S. 97 about 11 miles south of Oroville. Ice moving from left to right scraped the gently sloping steps, which face into the upstream direction of ice flow, and quarried the steeply sloping risers, which face in the down flow direction.

Following the Continental Suture

Between Brewster and Osoyoos, the highway follows the Okanogan River, along a line close to the boundary between the eastern edge of the North Cascade subcontinent and the trench filling that forms the bedrock in the Okanogan Valley, all that remains of the Okanogan trench. Hills in the distance west of the road are in the North Cascade subcontinent, those on the skyline east of the road in the Okanogan highlands. An ocean separated those hills on opposite sides of the road until the Okanogan trench closed it somewhere near the line of this road. It is fantastic to suppose that an ocean basin could utterly vanish, and leave no lasting souvenir of its existence beyond the rocks exposed in this area. But there is no simpler or more convincing explanation for the pattern of rocks on the geologic map of Washington.

Okanogan Dome

Rocks in the hills east of the road from Omak almost to the border belong to the Okanogan dome, an approximately circular mass of granite about 20 miles in diameter nearly surrounded by streaky and banded gneiss. Evidently, older crustal rock deep within the micro-continent partially melted to form a mass of magma that rose to become the granite core of the Okanogan dome. The rising granite magma dragged older rocks of the continental crust up with it, and they became the enclosing mantle of gneiss. Radioactive age dates on rocks of the Okanogan dome cluster around 50 million years.

Rocks in the near hills west of the road are far more varied. Very broadly speaking, they consist of deep oceanic sediments that were stuffed into the trench, and then invaded by several large masses of intrusive granite. The ages of the intrusive granites vary considerably. The older bodies tend to be about 90 million years old, the younger ones closer to 50 million years.

An Assortment of Lakes

Omak Lake, about 6 miles southeast of Omak, fills a bedrock basin in the floor of a large abandoned valley of a large stream. The origin of the abandoned valley is not completely clear. Some geologists suggest that the Columbia River may have flowed this way before the ice age glaciers diverted it to its present course. They may be right. There are many reasons to suspect that the drainage of this entire region flowed north before the ice age glaciers blocked the old streams.

Like several other small lakes in central and eastern Washington, Omak Lake has no outlet except through evaporation because the climate is too dry to keep the basin full enough to overflow. The lack of an outlet makes this one of Washington's least refreshing lakes. Its water contains enough salt and sodium carbonate to give it a distinctively bitter and soapy taste, but not enough to provide a profitable source of salt or sodium carbonate.

As its name suggests, Poison Lake is another extremely unpleasant body of water. Its water is a strong solution of magnesium sulfate, more familarly known as epsom salt, and a solid mass of crystalline epsom salt once lay beneath the lake. Most of that deposit was mined out during the first half of this century, first in an underground mine, later from an open pit. The material was purified in Tonasket, and shipped on from there.

Washington 20
Anacortes—Omak
203 mi. 324 km.

From One Coast to the Other

Between Anacortes on Fidalgo Island, and Omak, in the Okanogan Valley, Washington 20 carries the traveler from coast to coast of the old North Cascade micro-continent. The route crosses a bewildering complex of bedrock types. They include old continental crust, rocks stuffed into trenches on both sides of the subcontinent, and igneous rocks that formed because a trench exists off the western edge of the subcontinent. Most of the landscapes owe their origin primarily to glaciation.

Glaciation

The entire route passes through glaciated country. Anacortes and Sedro Woolley are in the flatlands of the Puget Sound lowland, which were deeply buried beneath the huge glacier that pushed south from Canada. Glacial sediments cover the low areas along that part of the route, and bedrock is exposed only in some of the hills.

West from Sedro Woolley to the pass, the highway follows the valley of the Skagit River, and then of its tributary, Granite Creek. An enormous glacier that started high in the mountains and flowed all the way to the Puget Sound lowland deeply filled this valley during the last ice age. Its effects are obvious in the relative straightness of long stretches of the Skagit River valley, and the steep valley walls. East of the pass, the road follows the valley of the Methow

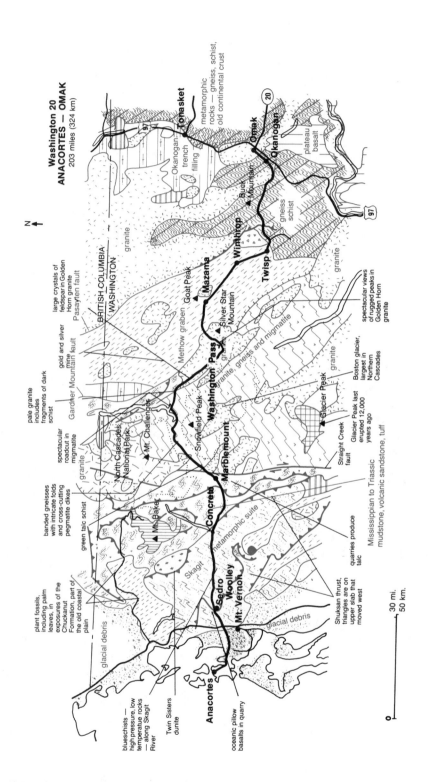

Washington 20
ANACORTES — OMAK
203 miles (324 km)

N

metamorphic
rocks — gneiss, schist,
old continental crust

Tonasket

Okanogan
trench
filling

Omak

Okanogan

plateau
basalt

Buck
Mountain

gneiss
schist

Winthrop

large crystals of
feldspar in Golden
Horn granite
Pasayten fault

granite

BRITISH COLUMBIA
WASHINGTON

Mazama

Goat Peak

Twisp

granite

gold and silver
mine
Gardner Mountain fault

Methow graben

Silver Star
Mountain

spectacular views
of rugged peaks in
Golden Horn
granite

pale granite
includes
fragments of dark
schist

spectacular
roadcut in
migmatite

North Cascades
National Park

Mt. Challenger

Snowfield Peak

Washington Pass

granite, gneiss and migmatite

Boston glacier,
largest in
Northern
Cascades

Glacier Peak

banded gneisses
with intricate folds
and cross-cutting
pegmatite dikes

granite

Marblemount

Straight Creek
fault

Glacier Peak last
erupted 12,000
years ago

Mississippian to Triassic
mudstone, volcanic sandstone, tuff

green talc schist

Mt. Baker

Concrete

Skagit
metamorphic suite

plant fossils,
including palm
leaves, in
exposures of the
Chuckanut
Formation, part of
the old coastal
plain

Sedro
Woolley
Mt. Vernon

quarries produce
talc

Shuksan thrust,
triangles are on
upper slab that
moved west

blueschists —
high pressure, low
temperature rocks
— along Skagit
River

Twin Sisters
dunite

glacial debris

glacial debris

Anacortes

oceanic pillow
basalts in quarry

30 mi.
50 km.

20

97

92

River, which is also uncommonly straight and steep walled because a glacier gouged it during the last ice age. That glacier joined the enormous mass of ice that filled the entire broad width of the Okanogan Valley.

The white slash of the North Cascades Highway cuts through forested slopes of the Methow Valley between Washington Pass and Mazama. The valley owes its broadly gouged form to a large glacier that scoured it during the last ice age.

West of the Straight Creek Fault

Washington 20 crosses the Straight Creek fault just east of Marblemount. Most of the rocks along the road between there and the Puget Sound lowland are dark gray mudstones and muddy sandstones that were stuffed into the trench that still exists off the west coast. In general, that trench filling becomes older and more severely deformed eastward.

The sinking ocean floor scraped the rocks in the eastern part of the area, between Concrete and Marblemount, into the trench while the region was still a micro-continent far offshore in the Pacific. The oldest of those rocks were laid down as oceanic sediments as much as 300 million years ago. Those older rocks were at one time deep enough in the trench to get very hot, and we now see them extensively recrystallized into metamorphic rocks. They rose to their present elevation as continued accumulation of the trench filling forced the ocean floor to sink along a more westerly line.

A striking pattern in a roadcut northeast of Marblemount. White dikes of coarse-grained granite pegmatite were injected as molten magma into much darker granodiorite, a rock related to granite.

Rocks between Concrete and the Sedro Woolley area did not get into the trench until about the time the micro-continent collided with North America. Except for being tightly folded and much broken along faults, those rocks still look like fairly ordinary oceanic sediments. However they are more thoroughly recrystallized than one might suspect from casual inspection, so geologists call them the Shuksan metamorphic suite. They include some interesting rocks.

Some of the rocks in the Shuksan metamorphic suite are blueschists. Close inspection of specimens collected from roadcuts will reveal occasional specks of sky blue in the generally dark gray mass of the rock. In a few places, the entire rock is blue. Blueschists form only where rocks were very rapidly stuffed into an oceanic trench, recrystallized under great pressure, and then returned to the surface before they could get very hot.

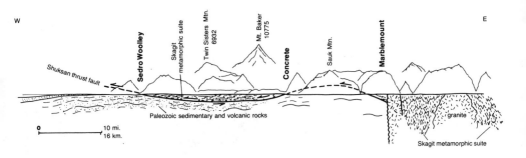

W E

Shuksan thrust fault

Sedro Woolley

Skagit metamorphic suite

Twin Sisters Mtn. 6932

Mt. Baker 10775

Concrete

Sauk Mtn.

Marblemount

Paleozoic sedimentary and volcanic rocks

granite

Skagit metamorphic suite

0 10 mi.
16 km.

Section approximately along the line of Washington 20 between Puget Sound and the Newhalem area. Geologists believe that all the rocks west of the Straight Creek fault lie on the Shuksan thrust fault, which dips east at an extremely gentle angle.

East of the Straight Creek Fault

Most of the rocks between the Straight Creek fault and the Mazama area belong to a complex of ancient igneous and metamorphic rocks, which geologists call the Skagit metamorphic suite. Most of those rocks are granite, or gneisses very much like granite. Gneisses have a streaky appearance because the mineral grains grew parallel to each other. Many gneisses also contain striking bands of light and dark rock. Even though gneiss and granite may look very much alike, geologists are careful to distinguish between them, because the aligned mineral grains in gneiss show that the rock recrystallized as it was squashed into a new shape. Many of the more homogeneous gneisses probably started out as granite, but those with bold color banding probably began as layered sedimentary rocks.

In some places along the highway, the layers in the banded gneiss swirl into disconnected marble cake patterns, and in other places intricate networks of white veins criss-cross the rock. Those are mixed igneous and metamorphic rocks called migmatite, which formed as the rock partially melted during metamorphism. The light part of the rock is granite, which was melted; the darker material is metamorphic rock, which remained solid.

Most of the granite and gneiss in the Skagit metamorphic suite high in the North Cascades is old continental crust that already existed when the micro-continent was far out in the Pacific. However, at least some of the granite formed after the micro-continent became part of North America. One of those granites, the Golden Horn batholith, appears along the road between the area just east of Marblemount and that several miles east of the Pass. Granite Creek is aptly named. The granite is a pale gray rock which consists mostly of milky white feldspar, along with gray grains of glassy looking quartz. Dark specks of hornblende or mica pepper the light mass of feldspar and quartz.

Ragged spires eroded in granitic rocks of the Golden Horn batholith watch over Washington 20 near Washington Pass.

The Methow Graben

Between the area a few miles west of Mazama and that about 5 miles east of Twisp, the road crosses the Methow graben. It lies between the Gardner Mountain fault, which is not visible in this area because it has vanished into younger granite, and the Pasayten fault. Unfortunately, the extremely interesting sedimentary rocks within the Methow graben are not exposed along the road because they weather easily, and lie beneath a deep fill of glacial debris.

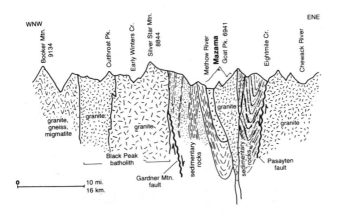

Section across the Methow graben. The sedimentary rocks that fill it are tightly folded, and contain some granitic intrusions.

If we could see the rocks in the Methow graben, we would find many of them full of the fossil remains of animals that lived in shallow sea water during Cretaceous and Jurassic time, as much as 150 million years ago. These formations consist of sediments that accumulated along a coastal plain, probably along the eastern edge of the North Cascade micro-continent while it was still far out in the Pacific.

The Okanogan Trench

Between the Pasayten fault about 5 miles east of Twisp and Omak, the road crosses another complex of igneous and metamorphic rocks. These appear to have formed in the Okanogan trench. However, exposure is extremely poor, because deep deposits of glacial debris bury most of the bedrock.

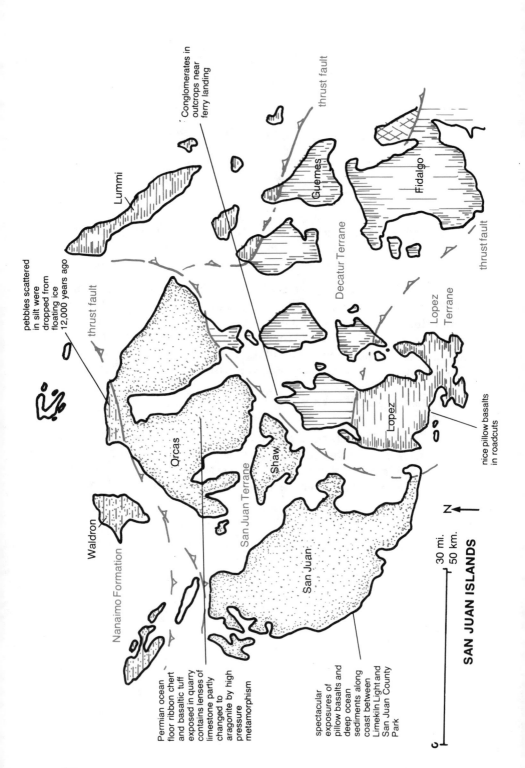

SAN JUAN ISLANDS

30 mi.
50 km.

N

thrust fault

thrust fault

thrust fault

Lummi

Guemes

Fidalgo

Decatur Terrane

Lopez Terrane

Lopez

San Juan Terrane

Orcas

Shaw

San Juan

Waldron

Nanaimo Formation

Conglomerates in outcrops near ferry landing

pebbles scattered in silt were dropped from floating ice 12,000 years ago

Permian ocean floor ribbon chert and basaltic tuff exposed in quarry contains lenses of limestone partly changed to aragonite by high pressure metamorphism

spectacular exposures of pillow basalts and deep ocean sediments along coast between Limekiln Light and San Juan County Park

nice pillow basalts in roadcuts

98

THE SAN JUAN ISLANDS

An Outpost of the North Cascades

At first glance, the San Juan Islands look like a geologic enigma, islands in the geologic map no less than in Puget Sound. However, they are merely a bridge between the rocks exposed in the North Cascades, and those along the nearby east side of Vancouver Island. The San Juan Islands contain rocks that were stuffed into an oceanic trench, probably during Jurassic time, about 150 million years ago. The San Juan Islands include large slabs of oceanic crust, along with severely deformed and recrystallized sedimentary rocks. Some of the metamorphosed sedimentary rocks are blueschists.

The Shuksan Fault

Some geologists suggest that the San Juan Islands were shoved west to their present position along the Shuksan thrust fault, which follows the eastern margin of the Puget Sound lowland. If so, then a deep well drilled anywhere in the islands should pass through the Shuksan fault and into younger rocks beneath, presumably into oceanic crust like that exposed in the northern Olympic Peninsula. If anyone ever discovers oil in commercial quantities in the northern Olympic Peninsula, it would become logical to drill such a well.

Waves breaking against Cattle Point, San Juan Island.

Three Terrains and Two Formations

Geologists divide the complex rocks of the San Juan Islands into three terrains in which the rocks appear to be related, even though varied, and two formations, which actually consist of one kind of rock. Although all three terrains consist of oceanic rocks that were stuffed into the trench, each contains distinctive assemblages of rocks, and it seems likely that faults separate them.

Old lime kilns at Roche Harbor at the north end of San Juan Island. Limestone suitable for making portland cement is scarce in western Washington, so that in the San Juan terrain was a valuable resource during the last century when transportation facilities were limited.

The San Juan terrain is the largest, and quite possibly the messiest. It contains a chaotic assortment of limestone that was deposited in shallow water, mudstone and chert that appear to have accumulated on the deep ocean floor, and scraps of oceanic crust. The rocks vary in age by at least 200 million years, and the only way they could now be together is through having been swept into a trench.

Veins of white plagiogranite intricately weave through black basalt of the oceanic crust exposed on Fidalgo Island. Plagiogranite is a curious rock, which typically occurs in small amounts in oceanic basalts. This material is part of the Decatur terrain.

The Decatur terrain is a bit more coherent. It seems to consist mostly of rocks formed during late Jurassic and early Cretaceous time, about 150 million years ago, in round numbers. Those rocks include thick sections of sandstones and pebble conglomerates scrambled in with slabs of oceanic crust. It seems reasonable to suspect that the sedimentary rocks may have been deposited on the edge of the oceanic crust. Then the sinking ocean floor stuffed both into the trench.

Basalt pillows of the oceanic crust exposed on Lopez Island.

The Lopez terrain also consists of rocks about 150 million years old, mostly sedimentary rocks that originally accumulated on the deep ocean floor. Deformation and heating in the depths of the trench converted most of the sedimentary rocks to dark slates, which break into thin slabs.

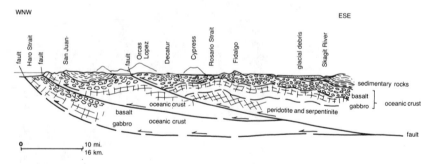

Section showing the San Juan Islands resting on the Shuksan and related thrust faults which moved them westward over younger rocks beneath.

The Nanaimo formation consists of much younger and much less deformed oceanic sediments exposed in several islands along the northwest edge of the group. It is similar to rocks exposed on the nearby coast of Vancouver Island, and along the north edge of the Olympic Peninsula. A deep well drilled on Orcas or San Juan Island should go through the thrust fault, and into the younger Nanaimo formation beneath.

102

IV
THE CASCADE VOLCANOES

Cascade volcanoes extend from Mount Garibaldi in southern British Columbia south through Washington and Oregon to Mount Lassen in northern California. They exist because the floor of the Pacific Ocean sinks beneath the west coast of North America from southern British Columbia to northern California. As the sinking lithospheric slab reaches a depth of about 60 miles, red hot steam rises to cause melting of the overlying already hot mantle to form basalt or andesite magma. Except for Mount St. Helens, which stands off to the west, the major active volcanoes lie along a remarkably smooth and continuous line. It marks where the sinking slab beneath gets hot enough to start the action.

Volcanic activity began in the North Cascade microcontinent long before it joined North America, and has continued ever since. But most of what geologists generally regard as the Cascade volcanic rocks erupted after the subcontinent docked. Two generations of volcanic rock exist in the Cascades: The older rocks of the Western Cascades, and the younger rocks of the High Cascades.

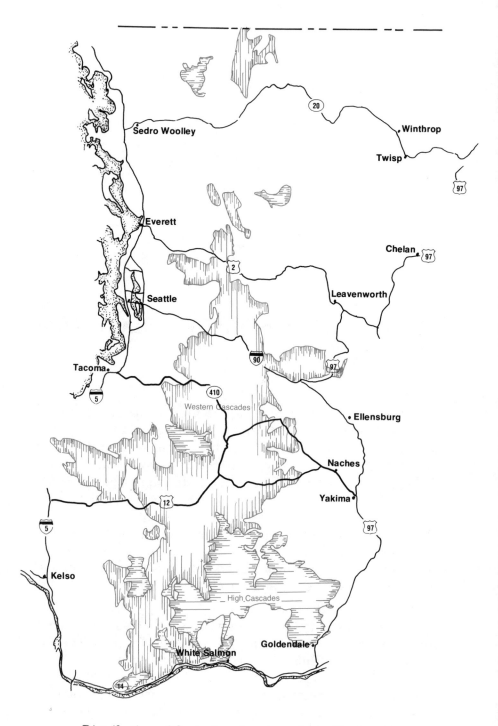

Distribution of Cascade volcanic rocks in Washington.

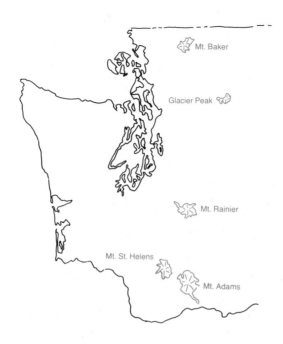

Major Peaks of the High Cascades.

The Western Cascades

There is no doubt that eruptions in the directly ancestral Western Cascade Range began in latest Eocene or early Oligocene time, approximately 40 million years ago. Geologists have traced those volcanic rocks in a continuous chain south into California, through regions of Oregon that are not part of the North Cascade subcontinent. As the name implies, those earlier Cascade volcanic rocks lie generally west of the modern High Cascades.

The Western Cascades remained active almost until eruptions began in the Columbia Plateau, until about 20 million years ago. That was time enough for many generations of large volcanoes to grow, erupt a wide variety of rocks, and then yield to new volcanoes. That long history of varied activity produced a complex chain of volcanoes, which would have been difficult to decipher if we could have seen them in their prime. Now, 20 million years of erosion have so carved those old volcanic edifices that it is rarely possible to recognize old volcanoes as individual mountains in the modern landscape of the Western Cascades. Instead, we see an ordinary erosional landscape carved into volcanic rocks.

BASALT, ANDESITE, AND RHYOLITE

Much of the lava that erupted from both ancient Western Cascade volcanoes and that now erupts from the modern Cascade volcanoes is common black basalt. Some of the lava is rhyolite, extremely pale rock that corresponds in composition to granite. However, most of the lava is andesite, a wonderfully broad name that applies to the range of gray volcanic rocks that fills the spectrum between basalt and rhyolite.

Basalt that erupts in the Cascades almost certainly comes from oceanic crust melting off the sinking slab beneath the region. Rhyolite forms most easily through partial melting of continental crust, in the Washington Cascades probably from melting deep in the North Cascade subcontinent. Andesite is intermediate in composition between basalt and rhyolite, and could form in several ways. Mixing of basalt and rhyolite magma could form andesite, although that seems unlikely to happen on a large scale. More complete melting of continental crustal rocks could also form andesite, as could contamination of basalt magma with small quantities of other kinds of rock. The differences in viscosity and stream content between basalt, andesite, and rhyolite explain a great deal of what volcanoes do.

Unless it contains steam, the only instigator of volcanic violence, molten rock behaves as quietly and as predictably as lukewarm oatmeal. Basalt and dark andesite magma contains very little steam, and is fairly fluid, so the steam can escape easily. Steam escaping from such magmas may cough chunks of flowing lava into the air as it escapes, and often blows off ominous clouds of dark ash. Those performances sometimes make impressive natural fireworks displays, but do not qualify as violent activity in the outrageous scale of volcanic standards.

Rhyolite and pale andesite magmas, on the other hand, absorb a great deal of water, and may arrive at the surface with a heavy charge of steam. Furthermore, those magmas are extremely viscous, so the steam can not bubble quietly out of the melt. Instead, the rock explodes. Nothing else in nature compares in violence with the explosion of several cubic miles of rock loaded with steam at a temperature somewhere above 750 degrees centigrade.

Basalt

Ropy surface of a basalt lava flow. The polygonal pattern in the background is a top view of the fractures that cause such flows to break into vertical columns in cliffs and roadcuts.

Basalt is perfectly black when fresh, although weathered surfaces quickly become red or brown, and reaction with water can alter basalt to various shades of dark green. Basalt melts at a temperature of about 1200 degrees centigrade, and erupts as lava about as runny as molasses, fluid enough to spread out thin into flows that cover large areas.

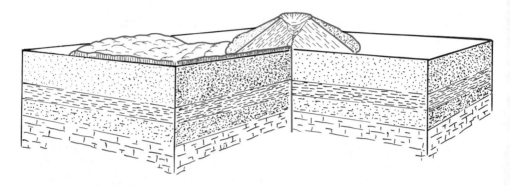

The anatomy of a cinder cone.

Most basalt or dark andesite eruptions begin as a long fissure, a crack several feet wide, opens in the earth's surface. The first magma to approach the surface commonly absorbs small amounts of water that blows off as steam through one or more vents along the line of the fissure, and coughs lava out of the vent. The smaller shreds of magma drift downwind as clouds of dark ash while the larger globs freeze as they fly through the air, and land near the vent as cinders of basalt. Cinders are bits and chunks of black rock as full of gas holes as fresh bread.

During several weeks or months, until the steam has escaped, cinders pile up into a small volcano called a cinder cone that typically grows to a height of a few hundred feet. Then the eruption ends as a large lava flow bursts out through the base of the loose pile of fragments. Most basalt and dark andesite flows have a rubbly surface, others a smooth ropy surface. Despite the difference in surfaces, the rock within is virtually the same.

Most cinder cones are single shot volcanoes that never erupt again. The next basalt eruption builds another little cinder cone, with a new lava flow. Cinder cones are extremely common in the Cascades. In some places they dot the landscape like a rash.

Sloping beds of basalt cinders exposed in a cut in a cinder cone.

U.S. Geological Survey photo

If basalt or dark andesite lava flows continue to erupt in the same area, they build a pile of thin lava flows that eventually grows into an enormous structure that geologists call a shield volcano. Actually, those volcanoes are shaped about like a gigantic vanilla wafer with a little crater in its top. Although the largest volcanoes in the world are basalt shields, most attract little attention because their sides slope too gently to add much emphasis to the landscape.

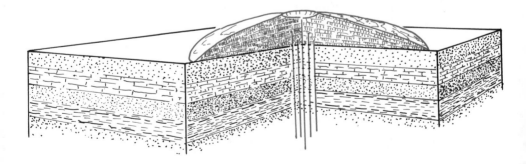

A cutaway view of a shield volcano.

Andesite

Andesites come in all shades of gray ranging from some that are nearly as dark as basalt to others that are almost as pale as rhyolite. Dark andesites resemble basalt in composition, steam content, and eruptive behavior. Paler varieties tend to erupt as viscous magmas which explode if they happen to carry a heavy charge of steam. Most of the andesite in the Washington Cascades is very dark, not too different from basalt.

Rubbly gray andesite agglomerate, the typical rock of a composite volcano.

Because they vary so greatly in composition, viscosity and water content, andesite lavas erupt in many ways that produce a corresponding variety of volcanic rocks. Explosive eruptions of pale andesite charged with steam produce clouds of ash and blanket the volcano with chunks of lava that may then move down the slope in mudflows of ash and assorted debris. Quieter eruptions of dark andesite produce lava flows. Together, those eruptions build the towering symmetrical cones that most people picture in their minds when they think of volcanoes.

Geologists call them composite volcanoes, because they contain such a disorderly assortment of lava flows, volcanic ash, mudflow deposits, and other rubble. It is impossible to predict what may happen when a big andesite volcano begins to erupt, because there is no way of knowing how viscous the lava will be, or how much water it may contain.

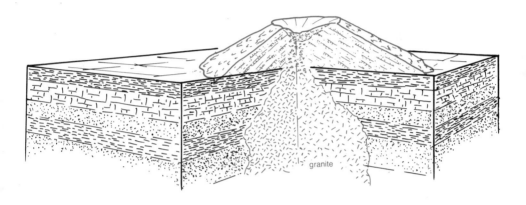

The structure of a composite volcano.

Rhyolite and Granite

Rhyolite is an extremely pale rock, generally almost white, or some weak pastel shade of yellow, pink, or green. It melts at a temperature of about 750 degrees centigrade, some 450 degrees lower than basalt. People who have watched rhyolite magma describe its consistency as resembling that of cold modeling clay. It seems more solid than fluid. A series of casual accidents determines whether a rising mass of rhyolite magma erupts to form a plug dome, explodes into a vast sheet of rhyolite ash, or crystallizes below the surface to become a granite pluton. It is difficult to think of kinds of rock that differ more in appearance, or less in substance.

If rhyolite or pale andesite magma absorbs little water, it remains molten as it approaches the surface, because the decreasing pressure progressively lowers its melting point. Under those conditions, rhyolitic magma erupts as very thick lava flows, or as bulging domes that typically swell into hills or small mountains over a period of several years. Geologists call such volcanoes plug domes or lava domes. Although they are extremely common in many parts of the Cascades, people tend to overlook plug domes because they look too much like ordinary small hills.

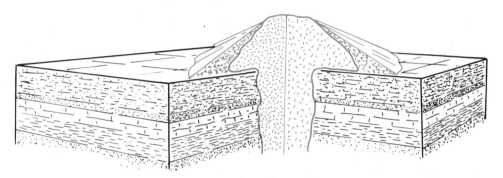

Cutaway view of a rhyolite plug dome.

If rhyolite or pale andesite magma arrives at the surface heavily charged with steam, it erupts in a mighty explosion that shreds the magma. Expanding steam puffs larger globs of magma into chunks so full of minute gas bubbles that they solidify into a rock foam called pumice that is typically light enough to float on water. Smaller bits of magma become volcanic ash, likewise full of minute gas bubbles. Most of the ash and pumice boils out of the volcano as a red hot ash flow that spreads at high speed over large areas, and then fuses into solid rock when it settles. The rest of the ash blows high into the sky to filter down over many thousands of square miles.

Chunks of white pumice littering the black surface of a basalt lava flow.

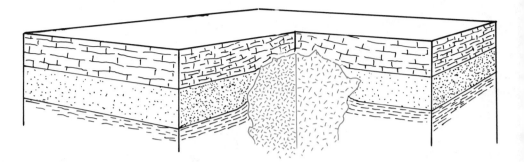

Structure of a typical pluton.

If rhyolite or pale andesite magma absorbs as much steam as it can hold, its melting point increases with decreasing pressure. Then, further rise to higher levels of the earth's crust will raise the melting temperature of the magma. If the magma rises high enough before it erupts to bring its melting temperature up to its actual temperature, it freezes. If that happens, the magma crystallizes below the surface to become a mass of granite that geologists call a pluton if it is small, a batholith if its outcrop extends over more than about 60 square miles. Most rising masses of rhyolitic magma become granitic plutons. Any volcanic chain that erupts rhyolite or andesite in its paler varieties also emplaces numerous granite plutons that remain to mark the spot long after erosion has stripped the volcanoes off the landscape.

MOUNT ST. HELENS

Before May 18, 1980, Mount St. Helens was the smallest, and to many eyes the loveliest, of Washington's big volcanoes. Its shimmering white cone, formerly 9677 feet high was so perfectly symmetrical that it looked almost the same from any direction. That simple conical form, unmarred by erosion, was evidence enough that Mount St. Helens was very young, therefore very active, almost every geologist's candidate for the Cascade volcano most likely to erupt soon.

An early blast of steam heralds the awakening of Mount St. Helens in March, 1980. At this time, the mountain was still perfectly symmetrical, an ideal example of a composite volcano.
U.S. Geological Survey photo

All of the other large volcanoes in Washington still bear the unrepaired scars of ice age glaciation. Obviously, they can not have been highly active during the past few thousand years. The absence of glacial sculpturing on Mount St. Helens means that the volcano has been extremely active since the last ice age. In fact, much of the volcano is less than 2000 years old and no rocks associated with Mount St. Helens are known to be more than 37,000 years old. Only a sustained high level of activity could build so large a volcano in so short a time.

Older rocks associated with Mount St. Helens show that the modern mountain stands on the remains of an extraordinarily violent predecessor that erupted large quantities of pale andesite. It covered much of the Pacific Northwest and Northern Rockies with ash. That earlier mountain probably consisted largely of a complex of lava domes that formed when masses of relatively dry lava erupted quietly. Violent eruptions happened whenever a mass of magma reached the surface with a heavy charge of steam.

Sometime around 2500 years ago, the volcano changed its habits, and began to erupt dark andesite and basalt, which ran out into thin lava flows. Those eruptions quickly built a shield volcano, the platform on which the modern volcano stands. Then, a long sequence of andesite eruptions built a conical edifice which looked perfectly symmetrical at a distance, but contained evidence of a complex history in the form of lava domes protruding from its flanks.

Throughout their complex histories, Mount St. Helens and its ancestor persistently produced large volumes of distinctly rhyolitic andesite. It is difficult to interpret that material as anything but partially melted continental crust. Therefore, it seems likely that Mount St. Helens stands on the North Cascade subcontinent, even though younger volcanic rocks completely cover any other evidence of that landmass in this part of Washington.

Spring, 1980, the Earliest Symptoms

No geologist interested in the Cascades was greatly surprised when Mount St Helens began to blow off steam on March 27, 1980, just 7 days after swarms of small earthquakes began to rattle windows in nearby communities. Large andesite volcanoes typically announce their eruptions with small earthquakes and clouds of steam. At first, the activity seemed fairly minor, modest bursts of white steam at a temperature hardly above the boiling point of water. It seemed possible that the steam was merely surface water that had soaked into the mountain, and found some hot rocks deep in its interior. That, after all, is what happened at Mount Baker in 1975.

Mount St. Helens wakes up. This plume of steam erupting in late April is almost white, evidence that the volcano was still producing little ash.—J.N. Moore photo

Earthquakes

All of the very small amounts of ash erupted during the early stages of the activity consisted of old rock, which contained no evidence of fresh magma newly arrived at the surface. However, Mount St. Helens was still generating numerous swarms of minor earthquakes, along with occasional tremors strong enough to shake a few dishes off the shelf. Volcanoes hardly ever cause really damaging earthquakes.

Geologists quickly established portable seismograph stations around Mount St. Helens to listen to the earthquakes. Those sensitive ears to the ground soon revealed that most of the earthquakes came from well below the volcano and shook only local areas of the surface. Steam rumbling inside the volcano could not cause that pattern of activity, but rising masses of magma typically do. Many of the earthquakes were harmonic tremors, which feel like a thrumming vibration instead of a series of sharp shocks. Harmonic tremors typically come from erupting volcanoes, evidently because magma moving up through the earth's crust causes them. They convinced most geologists that Mount St. Helens was indeed going into eruption, but told them nothing about what kind of activity to expect.

The Bulge

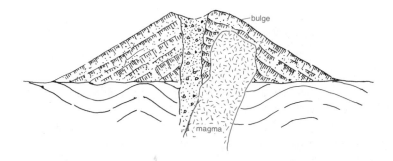

The situation within Mount St. Helens before the eruption of May 18, 1980.

The volcano continued to blow off clouds of steam that slowly blasted a new crater in the summit. Meanwhile, a bulge that had appeared on the north flank of Mount St. Helens when the activity started swelled at a steadily increasing rate. At first, only sensitive instruments could detect and measure the subtle swelling. But the tumor soon became ominously visible even to a casual observer as it grew faster, finally at a rate of about 6 feet per day. No one could doubt that a large mass of extremely viscous rhyolitic magma was intruding the north flank of the volcano. But that knowledge did not lead to a successful prediction because there were too many possibilities, and too many unknowns.

As so often happens with volcanoes, everything hinged on the question of water. If the mass of rising rhyolitic magma were dry, it would almost certainly extrude quietly through the flank of the volcano to produce another lava dome, or perhaps a thick tongue of a lava flow. Either would make an extremely tame eruption. On the other hand, if the rising magma were heavily charged with steam, it might explode — or it might crystallize quietly at depth without erupting, to form an intrusive granitic pluton.

Unfortunately, there is no way to measure the steam content of magma before it erupts, so there was no way to predict the course of events. Nevertheless, many geologists noticed that relatively little snow was melting off the bulge, and therefore suspected that the mass of magma rising into Mount St. Helens contained relatively little steam. A dome eruption seemed most likely.

The Triggering Landslide

Eventually, that swelling tumor on the north flank of Mount St. Helens grew large enough to pose hazards of its own. The slope had always been very steep, and now it was visibly growing steeper by the hour. At some time, it would certainly slide, and that might create a catastrophe, whether or not the volcano erupted.

Two geologists flying low in a light plane over the north flank of Mount St. Helens at 8:30 on the morning of May 18 noticed the snow melting rapidly from the bulge. That was an ominous sign, because it meant that steam was escaping from the magma beneath. Then, they saw the slope over the bulge begin to quiver. At the same moment, a small group of campers about 10 miles from Mount St. Helens happened to glance at the mountain and noticed that it looked blurred, as though it had somehow gone out of focus. Then, both parties saw the entire north slope of the mountain begin to slide. Fortunately, both took pictures.

Both sets of pictures show clouds of steam, dark with ash, begin to spurt out behind the moving slide. The pictures clearly show that the slide began to move before the eruption began. Within moments, the little jets of steam swelled to an awesome mushroom cloud.

May 18, 1980 – Mount St. Helens in full eruption.

The Explosion

Obviously, the invading mass of magma that raised the bulge on the north side of Mount St. Helens did indeed contain a heavy charge of steam. The weight of the rock above the magma exerted enough pressure to keep the lid on until the morning of May 18. Then the slide moved, uncorking the magma, and the steam exploded. Remember that we are not discussing ordinary steam of the tea kettle variety. Granitic magmas exist at temperatures above 750 centigrade, a bright red heat. Steam venting directly from such magma is hot enough to glow in the dark as a flame, hot enough to ignite wood upon contact. It is the steam that some magmas contain, not the molten rock itself, that inspires volcanoes to violence.

The steam explosion of May 18 released energy equivalent to 21,000 atomic bombs like the one that devasted Hiroshima. It shot the bulge horizontally northward in a cloud of ash and rock moving about 500 miles per hour. Meanwhile, the sliding north slope of Mount St. Helens moved down the mountain at a speed of about 200 miles per hour into the North Fork of the Toutle River. It continued downstream as a mud flow into the Columbia River, where it filled the navigation channels. The volume of that slide was about two thirds of a cubic mile.

The Toutle River choked with debris after the eruption of May 18, 1980. The picture has low contrast because gray volcanic ash covers everything. U.S. Geological Survey photo

The Ash Fall

During all that morning and most of the afternoon, the volcano continued to blow off its dark cloud of hot steam and ash. Only part of the ash blew into the air; most boiled out of the crater, and poured across the ground surface as a scorching ash flow that covered a swath all the way to Spirit Lake, about 5 miles, to an average depth of approximately 100 feet. The cloud of airborne ash drifted east, blackening the sky of a brilliant early summer day. It reached Spokane by mid-afternoon, and then continued across northern Idaho and western Montana during the late afternoon and evening.

Before the sun went down behind a cloud darker than anyone in the region had imagined possible, ash was falling on about one half million square miles of three states. It fell like snow through a windless night. The next morning, the sun rose through skies still gray with ash to shine wanly on a landscape drained of all color. Every surface was dusted with medium gray ash about the color and consistency of dry portland cement freshly shaken from the bag. Three days later, the sky was still ashen, three months later clouds of ash still puffed out of the trees when the wind blew, and three years later a coat of ash still stubbornly painted rock surfaces across a wide area of central and eastern Washington.

The Dome Eruption

When the sky cleared, Mount St. Helens was 1300 feet lower than it had been, and much less shapely. The landslide and explosion had opened a crater in its north flank right where the bulge had been, a yawning amphitheater about a mile across. The next phase of the eruption was already beginning.

The explosion of May 18 blew the steam-charged top off the rising column of magma, leaving intact the lower part, which contained much less steam. The magma continued to rise, just as it had been rising before the explosion, to form a dome in the floor of the new crater. Such domes typically grow at a diminishing rate for several years before they finally stabilize. Occasional steam explosions, which happen at increasing intervals, and with decreasing violence generally accompany the growth of a lava dome. They happen because the magma within the dome is cooling and crystallizing.

Most of the minerals that crystallize from magmas contain no water. Therefore, as crystals growing within a magma occupy more space, they crowd the steam into a diminishing

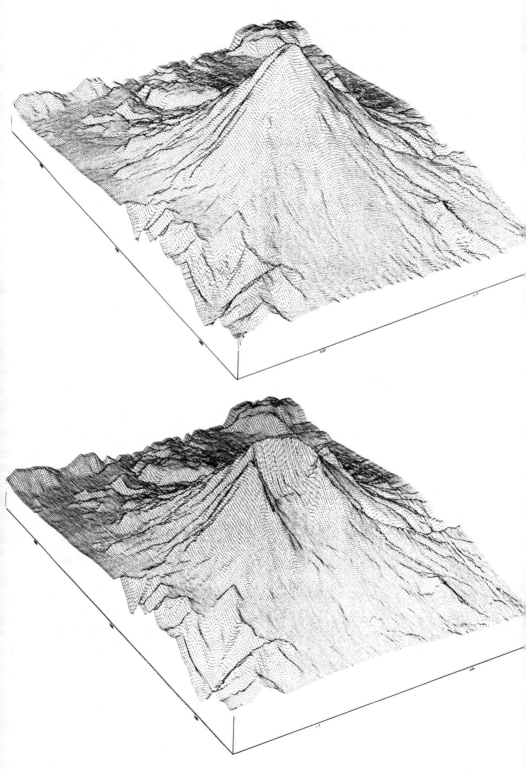

Mount St. Helens—before and after.—U.S. Geological Survey photo

volume of still molten magma. As the space available to the steam decreases, its pressure must increase. Air pressure within a rubber balloon increases as you squeeze it for the same reason. Eventually, the steam within the magma develops enough pressure to explode, just as the squeezed balloon eventually bursts. Occasional steam eruptions due to magma crystallization may continue for years, but they are not likely to repeat the performance of May 18, 1980.

The Future

It is impossible to know now whether the eruption of 1980 was an isolated event, or the beginning of a series of eruptions. However, the recent history of Mount St. Helens has been one of eruptive episodes lasting at least several decades separated by much longer periods of quiet. The last eruption before that of 1980 happened in 1857. It ended a series of at last 5 significant eruptions, several of historic record, that began about 1800 and produced lava flows, a lava dome and clouds of ash.

There is no historic record of earlier spells of activity, so our information comes mostly from radiocarbon dates on wood charred by lava flows or ash deposits. Evidently, there was a series of eruptions during the fifteenth and early sixteenth centuries quite similar to that of the nineteenth century. A still earlier, and also generally similar, series of eruptions happened about 700 years before that. And so it goes, periods of activity lasting several decades separating quiet intervals that last several centuries.

No one knows why that pattern of intermittent spells of intense activity prevailed in the past, or whether it will continue, although nothing suggests that the mountain has changed its ways. If Mount St. Helens continues to behave as it has during the past several thousand years, then the eruption of 1980 will probably prove to have been the first in a series that will continue well into the next century. Most geologists find that a fascinating prospect.

The Volcanic Perspective

By any ordinary reckoning, the Mount St. Helens eruption of May 18, 1980 was an extraordinary event. On average, comparable eruptions happen somewhere in the world only once in every two decades. However, it was not especially impressive by volcanic standards. The total volume of material erupted on May, 18, 1980 amounted to less than one cubic mile, considerably less than one tenth that erupted when Crater Lake, Oregon formed about 6200 years ago. The ash flows extend only about 5 miles from Mount St. Helens; geologists have found many ash flows that continue for more than 10 times that distance. It could so easily have been far worse.

MOUNT ADAMS

Mount Adams, 12,307 feet high, is second in height and bulk to Mount Rainier among Washington volcanoes, an enormous composite volcano by any standard. Its hulking mass stands about 50 miles east of Mount St. Helens, a solitary white sentinel that watches silently over an immense area of southern Washington. Despite its enormous bulk and commanding presence, Mount Adams is the least known of the large Cascade volcanoes. Its remote location and the ruggedness of the surrounding countryside make Mount Adams difficult to visit. There are no thorough geologic studies.

Mount Adams is not only considerably larger than its formerly more shapely neighbor to the west, but much different in general appearance. Its broadly squat form, which looks distinctively different from every direction, suggests that Mount Adams may actually be a congregation of several coalescing volcanic cones. Certainly that irregular and asymmetrical form records a long and complex volcanic history.

Mount Adams stands on a broad platform of basalt that partly explains its sprawling profile. Evidently, a long series of basalt eruptions built a broad shield volcano before eruptions of more viscous andesite began to build the towering cone. Most of the large composite volcanoes in Oregon and California developed in that way.

Glaciation

Unlike Mount St. Helens, the deeply furrowed sides of Mount Adams bear the scars of thousands of years of glaciation, and large glaciers actively gouge the mountain today. The Klickitat Glacier on the east side of Mount Adams is the second largest in the Cascade chain. All that glacial sculpturing is clear evidence that Mount Adams has not been significantly active for a long time, quite possibly not since the last ice age ended about 10,000 years ago. However, there have been small eruptions since the last ice age — at least one lava flow may be no more than a few thousand years old. And there are steam vents near the summit. So Mount Adams is still alive, even if moribund.

The Sulfur Mine

Most active volcanoes emit sulfurous gases. People who climb them on a quiet day often find the atmosphere near the summit anything but refreshing. Nevertheless, Mount Adams is the only large Cascade volcano to contain large deposits of sulfur in its crater. They were mined on a small scale for many years, but the cost of operating in that remote region finally made the operation hopelessly unprofitable. A mine in the top of Mount Adams can not compete with those along the coast of Louisiana and Texas, which inexpensively produce large quantities of extremely pure sulfur.

Future Prospects

So far as anyone knows, Mount Adams consists almost entirely of dark andesite, and has no history of erupting rhyolitic lavas. Therefore, another eruption, if there is another eruption, probably will produce more andesite, not a big rhyolite explosion. Nevertheless, there are hazards. Like all large andesite volcanoes, Mount Adams is likely to produce mudflows for many thousands of years, even if it never erupts again. Composite volcanoes are simply too steep, and the rocks in them too weak, to have stable slopes. However, the area within mudflow range of Mount Adams is virtually uninhabited, so any mudflows that may pour down the mountain pose very little threat to human life or property. Of all the big volcanoes, Mount Adams probably comes closest to being all scenery, and no hazard.

MOUNT RAINIER

Mount Rainier, 14,410 feet almost directly from sea level, is one of those astounding mountains that makes you crane your head back to see its summit. Its perpetually white peak, clad in big glaciers, 26 of them, looms over the landscape of much of western Washington. No one can live within Rainier's domain without feeling the brooding presence of that emphatic punctuation mark on the skyline.

Nisqually Glacier scours the south flank of Mount Rainier.

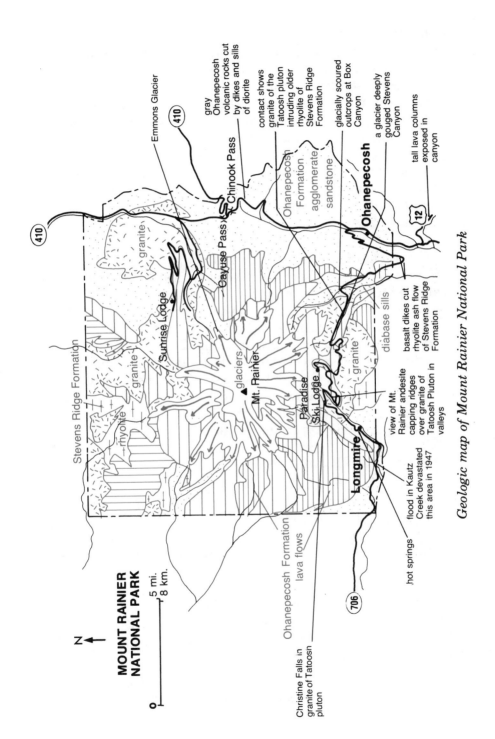

Geologic map of Mount Rainier National Park

MOUNT RAINIER NATIONAL PARK

N

0
5 mi.
8 km.

Emmons Glacier

gray Ohanepecosh volcanic rocks cut by dikes and sills of diorite

contact shows granite of the Tatoosh pluton intruding older rhyolite of Stevens Ridge Formation

glacially scoured outcrops at Box Canyon

a glacier deeply gouged Stevens Canyon

tall lava columns exposed in canyon

Chinook Pass

Ohanepecosh Formation agglomerate sandstone

Ohanepecosh

410

410

Cayuse Pass

granite

Stevens Ridge Formation

Sunrise Lodge

granite

rhyolite

glaciers

Mt. Rainier

granite

Ohanepecosh Formation lava flows

Paradise Ski Lodge

diabase sills

basalt dikes cut rhyolite ash flow of Stevens Ridge Formation

12

Longmire

view of Mt. Rainier andesite capping ridges over granite of Tatoosh Pluton in valleys

flood in Kautz Creek devastated this area in 1947

hot springs

Christine Falls in granite of Tatoosh pluton

706

127

Ancient History

Rocks exposed around Mount Rainier show that it stands on the eroded wreckage of older volcanoes of the Western Cascades that began forming at least 40 million years ago. The oldest of those rocks comprise the Ohanepecosh formation, a thick series of very dark andesite lava flows, and light-colored sedimentary rocks composed of debris eroded from them. A few of the sedimentary rocks contain fossil plants, which include palms and other trees that must have required a climate considerably milder than the one we know today. After the Ohanepecosh volcanoes snuffed out, the rocks they erupted were folded, and a rugged landscape eroded into them before the next phase of volcanic activity began.

The second episode of volcanic activity produced light colored volcanic rocks, mostly rhyolite volcanic ash, which geologists call the Stevens Ridge formation. Those rocks filled the valleys and covered the hills eroded into the Ohanepecosh formation, building a new volcano of which very little remains. Then, further volcanic activity produced the Fifes Peak formation, another sequence of very dark andesite lava flows and related sedimentary rocks. We can be sure that both the Stevens Ridge and Fifes Peak volcanoes erupted before late Miocene time, because flood basalt flows of the Columbia Plateau lap over them in the area east of Mount Rainier.

The Tatoosh Pluton

Next, a large mass of granitic magma invaded the area intruding all the older rocks, and erupting a new sequence of light colored volcanic rocks, more rhyolite. As usual with rhyolitic volcanic rocks, at least some of the eruptions appear to have been explosive. As the magma continued to rise, it finally intruded its own cover of volcanic rocks, and then crystallized as a mass of granite called the Tatoosh pluton. Now that streams and glaciers have cut valleys into the complex, geologists can trace the light colored volcanic rocks down into the body of granite that erupted them before it crystallized.

The Tatoosh granite pluton and the various rhyolitic volcanic rocks suggest that the North Cascade subcontinent probably lies buried beneath the younger volcanic rocks in this region. Granitic magma, which becomes rhyolite if it erupts, forms most easily through melting of continental crustal rocks. Andesite volcanoes that stand on oceanic crust generally do not erupt rhyolite.

An Interlude

Emplacement of the Tatoosh pluton seems to have been the last stage in the first long series of igneous events in the Mount Rainier area. Then this area, like all of the Cascade Range, became volcanically inactive as the focus of activity shifted east to the Columbia Plateau. Streams carved the older rocks into a rugged landscape in which the ridge crests stood several thousand feet above the valley floors.

Mount Rainier

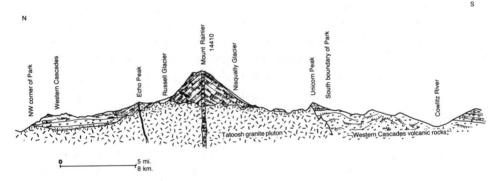

Section through Mount Rainier showing the modern volcano standing on an erosion surface on the older Tatoosh granite pluton.

No one knows exactly when renewed eruptions began to build the Mount Rainier we know, but it must have been sometime around one million years ago. The first eruptions filled deep canyons in the old landscape with lava, and then the mountain grew as lava continued to rise through the same vent. Now the earliest canyon-filling lava flows stand up in relief because they resist erosion more effectively than the surrounding rocks. Those oldest flows form ridges with curves that record the windings of the old stream valleys, and radiate from Mount Rainier like big roots around an old stump.

129

Virtually all of Mount Rainier consists of very dark andesite. Like many large composite volcanoes, Rainier began its career by erupting dark and fluid lavas that spread over large areas to build a broad base, essentially a shield volcano, which in this case covers about 100 square miles. Volcanic debris in the Puget Sound lowland suggests that early stages of Mount Rainier's growth may also have involved a number of explosive eruptions. However, most of the exposed rocks in Mount Rainier are lava flows that must have erupted fairly quietly, even though they were viscous enough to pile up around the crater to build the towering composite cone. Mount Rainier also contains enough cinders and ash to assure us that at least some of its eruptions, although not violent by volcanic standards, produced spectacular fireworks displays.

Glaciation

Mount Rainier has probably worn its mantle of snow and ice ever since it grew high enough to catch the clouds. Quite a few of those dark andesite lava flows must have erupted under glaciers, or flowed onto glaciers. Any encounter between a lava flow and a glacier produces spectacular floods and catastrophic mudflows, along with all sorts of peculiar pyrotechnics. One lava flow, now exposed in Ptarmigan Ridge on the north side of Mount Rainier, consumed a glacier more than 1000 feet thick, and filled its valley with andesite.

Fine silt and rocks embedded in the sole of an ice age glacier polished and grooved this rock surface in the Box Canyon of the Cowlitz, Mount Rainier National Park.

Large glaciers of the last ice age carved Mount Rainier so deeply that little remains of its original volcanic form. The mountain is no longer conical, and is so craggy that it looks different from every vantage point. Furthermore, nothing remains of the original volcanic surface — Rainier is much smaller now than it was when it reached its maximum size sometime around 75,000 years ago. All the deep glacial sculpture clearly shows that Mount Rainier has not been active enough to keep pace with erosion for many thousands of years.

The Most Recent Activity

The only solid geological evidence of significant activity in the last several thousand years is a small summit cone, a new and quite shapely little volcano about 1000 feet high on the summit of Mount Rainier. Two small craters in its top are filled with snow. The series of eruptions that built that cone sent an avalanche of hot rock pouring down the Puyallup Valley where carbonized trees have yielded radiocarbon dates of about 2500 years ago. The same eruptions produced andesite lava flows that melted large parts of the Emmons and Nisqually glaciers, thus causing enormous mudflows.

The last few centuries have brought a number of extremely minor eruptions on Mount Rainier. One brief episode of activity produced some pumice between 500 and 600 years ago, to judge from the ages of trees killed then, and those that have grown since. Another minor outburst scattered pumice over small areas on the east side of the volcano sometime during the first half of the last century. Although the newspapers of a century or more ago contain numerous accounts of eruptions on Mount Rainier, little geologic evidence in the form of fresh lava flows or ash beds confirms those reports. Indeed, the old newspapermen were notoriously unreliable volcano watchers who saw eruptions everywhere, even on peaks that are not volcanoes, such as Mount Olympus. It is impossible now to know what forest fires and strange clouds they saw, but more recent events suggest that some of their accounts may have been factual.

Reliable reports with solid geologic evidence behind them show that Mount Rainier produced a number of respectable

This part of Emmons Glacier is black because rock fall debris covers the ice.

steam explosions during the 1960's and 1970's. One such explosion may have triggered the big rockfall of December, 1963, which broke some 14 million cubic yards of rock off Little Tehama Peak, and dumped it onto the Emmons Glacier. Most of the rock slid down the glacier, off its lower end, and about 4 miles on down the valley. The mass of broken rubble landed near the White River Campground, and contains blocks as big as houses.

Will Mount Rainier Erupt?

Although Mount Rainier has been relatively inactive for many thousands of years, there is no reason to suppose that it is extinct. Even quite a few thousand years is not a long time in the life of a volcano so old and so large. So we have to consider the possibility of future eruptions of Mount Rainier, and there is no way to predict what might happen.

Most of the lava in Mount Rainier is very dark andesite, which typically erupts quietly, and poses almost no hazard of great violence. However, even a modest flow of the quietest dark andesite can raise havoc in the form of devastating floods and mudflows if it happens to melt large quantities of snow and ice. And some very large composite volcanoes do end their careers with a cataclysmic eruption, an armageddon of pumice and ash in which the mountain subsides into a crater as cubic miles of rhyolite erupt from beneath it. Oregon's Mount Mazama, which was about the size of Rainier, did that 6200 years ago, leaving nothing of itself but the circular rim that now encloses Crater Lake.

Granite of the Tatoosh pluton and rhyolite of the Stevens Ridge formation both show that large masses of rhyolite magma have appeared at the site of Mount Rainier in the past, and so could again. Furthermore, Mount Rainier has erupted small quantities of rhyolite pumice during the geologically recent past. We can not discount the possibility of a catastrophic rhyolite eruption.

Mudflows

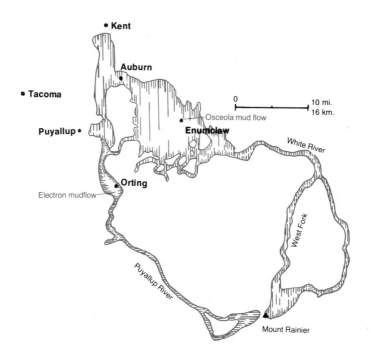

The Osceola and Electron mudflows.

Regardless whether it erupts again, Mount Rainier will certainly continue to produce large mudflows for thousands of years. The Osceola mudflow, for example, happened without benefit of an eruption about 5000 years ago. It started when a large part of the peak of Mount Rainier collapsed, and then poured northwest down the White River valley. That mudflow contains about one-half of a cubic mile of debris, and must have made an appalling wall of mud as it spread across more than 100 square miles of the Puget Sound lowland near Enumclaw. A modern repetition of the Osceola mudflow would kill many thousands of people, and cause hundreds of millions of dollars in property damage. Mudflows are too big to stop, and most of them move too fast to permit people to run away.

Many lesser mudflows have poured off Mount Rainier during the last few thousand years. The Electron mudflow, for example, poured down the valley of the Puyallup River almost to Sumner, burying everything in its path within a few hours. There is every reason to expect more mudflows in years to come. Rainier is still a dangerous neighbor.

GLACIER PEAK

The gnarled pinnacle of Glacier Peak rises 10,528 feet into the skyline east of Seattle. Although it makes an emphatic point on the ridge of the Cascades, Glacier Peak lacks the volume of the other High Cascade volcanoes. It gains much of its elevation from its perch on a high ridge composed partly of ancient gneiss, basement rock of the North Cascade subcontinent. Like the other big Cascade volcanoes, Glacier Peak has a long and complicated history of igneous activity.

Early History

A mass of granite, the Cloudy Pass batholith, lies beneath large areas around Glacier Peak, presumably beneath the volcano as well. That mass of granite magma invaded the very much older basement rock of the North Cascade subcontinent about 22 million years ago, during Miocene time. It crystallized at very shallow depth, after having violently erupted part of its volume.

134

Volcanic rocks on Gamma Ridge, east of Glacier Peak, appear to have erupted from the same granite magma that became the Cloudy Pass batholith. Those rocks include rhyolite ash, welded ash, and pumice, all typical products of violent eruption of granitic magma, which has the composition of rhyolite. Furthermore, they are the right age, and they contain fragments of granite identical to that in the Cloudy Pass batholith.

The Modern Volcano

A long period of quiet that must have lasted at least 10 million years followed formation of the Cloudy Pass batholith and its companion, the Gamma Ridge volcanic pile. Then renewed eruptions began to build the Glacier Peak volcano that we see today on the same site. No one knows exactly how old the Glacier Peak volcano is, but we can be sure that it began to grow less than 700,000 years ago. That figure emerges from study of the magnetism of the oldest exposed lava flows.

The earth's magnetic field imprints itself on most kinds of rocks as they form, and they retain a permanent record of the direction in which a compass needle would have pointed at the time and place where they formed. For reasons that no one understands, the earth's magnetic field reverses from time to time — the north and south magnetic poles simply trade places. On average, that happens about every half million years, although the individual periods vary greatly. All the lava flows in Glacier Peak are magnetically polarized in the same direction as the present magnetic poles, so it seems that all must have formed since the last magnetic reversal, which happened about 700,000 years ago.

During most of its history, Glacier Peak erupted lava flows of dark andesite quite unlike the pale rhyolitic rocks of the older Gamma Ridge volcano. The relatively quiet eruptions of fluid andesite lava built a volcano consisting mostly of lava flows. Rather late in its career, Glacier Peak produced a large dome of viscous andesite that bulged out of the volcano and oozed heavily down its slope. That dome is now prominently visible as Disappointment Peak, a satellite summit on the south flank of the volcano.

A Rhyolite Explosion

During the most recent, and perhaps the final, eruption of Glacier Peak, the volcano changed its habits and explosively erupted a large mass of rhyolite. The ash from that eruption covers much of Glacier Peak, and therefore appears to have erupted from the summit of the volcano. The same ash also blanketed much of the Pacific Northwest, including northern Idaho and western Montana. The event has been accurately dated at about 12,000 years ago by radiocarbon analysis of charcoal incorporated in the ash. Everywhere it occurs, the Glacier Peak ash forms a distinctive pale layer in the soil. It is widely used for dating a variety of things ranging from glacial moraines to archeological sites.

The eruption of 12,000 years ago came after a long period of relative quiet, and another such period has followed it. A volcano that has not erupted for 12,000 years, and that did not erupt for a long time before then, may well be extinct. Furthermore, many large andesite volcanoes end their careers with a final glorious eruption of pale rhyolite. At the very least, we are safe in concluding that the recent history of Glacier Peak suggests that it is now erupting at very widely spaced intervals. Of the five Cascade giants in Washington, Glacier Peak seems the one least likely to erupt again.

MOUNT BAKER

Mount Baker, 10,778 feet high, watches over the northern end of the Cascade chain. Its striking cone clad in sparkling white glaciers, about 20 square miles of them, dominates the skyline of a vast area, a silent presence in the lives of many thousands of people. Enormous glaciers of the last ice age, vastly larger than those that now cover the top of the mountain, gouged deep valleys in Mount Baker. They left the volcano gnarled and craggy, with little of its original conical form or volcanic surface intact. That deeply eroded shape clearly shows that Mount Baker has not erupted very much for a long time. However, even though Mount Baker is not active enough to repair the scars of ice age glaciation, it certainly is not extinct.

Mount Baker resembles its much larger neighbor, Mount Rainier, in consisting almost entirely of very dark andesite. The mountain stands on a base of very thick lava flows that filled old valleys, and consists mostly of lava flows with relatively little fragmental material. To judge from the rocks, most of the eruptions that built Mount Baker must have been fairly quiet.

Thin layers of ash and several lava flows on the slopes of the volcano provide positive evidence of several eruptions since the last ice age. The largest eruption since the last ice age built a basalt cinder cone at the base of Mount Baker, and produced a lava flow that went some 12 miles down the valley of Sulfur Creek. Numerous historical accounts, some of them fairly convincing, tell of several eruptions between 1843 and 1859. Those reports all describe clouds of ash and steam issuing from the Sherman Crater about one-half mile south of the summit. However, none of those events left significant deposits of new volcanic rock on the slopes of the volcano.

The False Alarm of 1975

Mount Baker constantly vents small quantities of steam, mostly from the Sherman Crater. It often forms a small cloud over the peak. During March of 1975, Mount Baker began to blow off impressive clouds of steam in a striking white plume that was visible for many miles. At times, small quantities of ash darkened the plume, and for a while the volcano produced more than a ton of sulfurous gases every hour. It seemed possible that Mount Baker might go into full eruption, thus becoming the first Cascade volcano to show much sign of life since 1917, when Mount Lassen staged an impressive performance in California. The U.S. Geological Survey immediately sent a team of geologists to Mount Baker to watch the activity, and attempt to predict its course.

The survey geologists quickly found that the temperature of the steam was about at the boiling point of water, too cool to suggest that it was coming from magma. When volcanoes are ready to erupt, they typically produce steam at more than twice the boiling point of water. And the clouds of steam were almost white, which meant that they were carrying very little volcanic ash. Furthermore, microscopic examination revealed that samples of the ash contained only fragments of old volcanic rock, no freshly erupted material to show that a mass of magma was rising in the volcano. Seismograph monitoring for earthquakes recorded only one small tremor in the next year, probably less than the usual number of earthquakes in the vicinity of Mount Baker. In short, Mount Baker showed no symptom of imminent eruption.

Evidently, Mount Baker was simply blowing off surface water that had soaked into the mountain, and come into contact with hot rocks in its interior. Clouds of steam continued to appear for several months, and then tapered off fairly rapidly. Two years later, Mount Baker was once again emitting its customary small plume from the steam vents in the Sherman Crater.

Mudflows

Whether or not Mount Baker erupts again, it is still a potential source of extremely dangerous mudflows. Large andesite volcanoes consist mostly of broken rock, including large amounts of volcanic ash, that stand precariously on very steep slopes. Numerous mudflow deposits around Mount Baker are clearly less than 10,000 years old because they cover the floors of valleys glaciated during the last ice age. Many must have come down the mountain since its last major eruption. One very large mudflow poured almost 20 miles down the valley of the Nooksack River about 6000 years ago, burying everything in its path. Other mudflows have moved southeast down Boulder Creek towards Baker Lake during the last 10,000 years. That valley heads near the Sherman Crater where numerous steam vents melt ice and snow, thus providing water for the mudflow mixture.

I-90
Seattle—Ellensburg
107 mi. 170 km.

Between Seattle and Ellensburg, Interstate 90 crosses the Cascade Range from the Puget Sound lowland in the west, to the edge of the Columbia Plateau in the east. The road goes from one former coast of the North Cascade subcontinent to the other, crossing some rocks that were once part of an island far offshore, and others that formed since the island became part of North America. The rocks and landscapes along the way tell several geologic stories, most obviously that of ice age glaciation.

Glaciation

The highway follows the Snoqualmie River on the western slope of the Cascades to the divide at Snoqualmie Pass, the Yakima River from the pass to Ellensburg on the eastern slope. Large glaciers filled both valleys during the last ice age, so the road crosses deep deposits of glacial debris that fill the valley floors along most of the route.

The glacier that filled the Snoqualmie Valley descended almost to

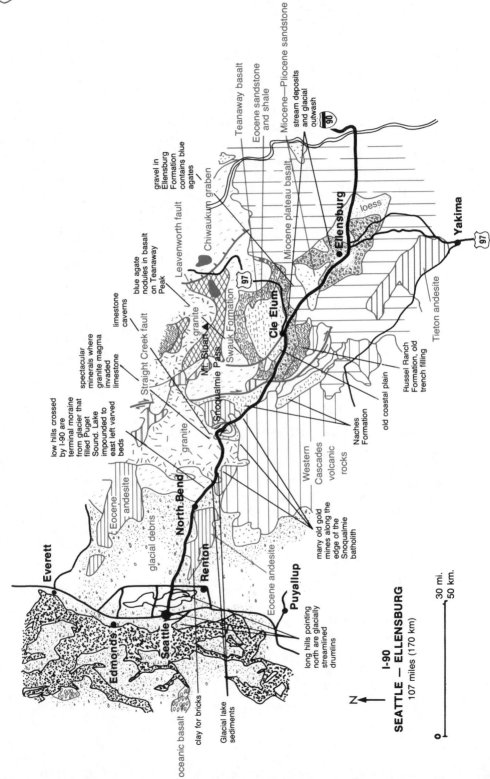

I-90
SEATTLE — ELLENSBURG
107 miles (170 km)

N

0   30 mi.
     50 km.

Everett

Edmonds

Seattle

Renton

Puyallup

North Bend

Cle Elum

Ellensburg

Yakima

oceanic basalt

clay for bricks

Glacial lake sediments

long hills pointing north are glacially streamlined drumlins

Eocene andesite

glacial debris

granite

granite

Eocene andesite

Straight Creek fault

Snoqualmie Pass

Mt. Stuart

Swauk Formation

many old gold mines along the edge of the Snoqualmie batholith

Western Cascades volcanic rocks

Naches Formation

old coastal plain

Russel Ranch Formation, old trench filling

Tieton andesite

loess

Miocene plateau basalt

spectacular minerals where granite magma invaded limestone

limestone caverns

blue agate nodules in basalt on Teanaway Peak

low hills crossed by I-90 are terminal moraine from glacier that filled Puget Sound. Lake impounded to east left varved beds

Leavenworth fault

Chiwaukum graben

gravel in Ellensburg Formation contains blue agates

Teanaway basalt

Eocene sandstone and shale

Miocene–Pliocene sandstone

stream deposits and glacial outwash

90

97

97

90

The deep notch in the mountainside north of I-90 just east of North Bend is an ice marginal stream channel. A large meltwater stream flowing between the mountainside and the ice that filled Puget Sound carved the channel during the waning stages of the last ice age, perhaps about 12,000 years ago.

sea level, where it joined the expanse of ice that filled the Puget Sound lowland. The glacier that flowed east down the valley of the Yakima River ended at Thorp Prairie 13 miles east of Cle Elum, leaving a large moraine there to record its farthest reach. Both glaciers left their plainest mark in the broadly gouged valleys with flat floors deeply filled with debris left as the ice melted. Those valleys are much too straight to have been eroded by streams alone, much too wide to fit the streams that now flow through them. And the steep valley walls also tell of ice age glaciers.

Coal

Between Seattle and the Preston area, the highway passes just north of the King County coal fields. The coal seams are in sedimentary formations which geologists lump into the Puget group. The original sediments, mostly sand and mud, were deposited along the western coastal plain of the North Cascade subcontinent during Eocene time, approximately 50 million years ago. That was about the time that the micro-continent joined North America. The coal beds almost certainly formed from peat laid down in tidewater swamps along the old coastline. Unfortunately, glacial deposits cover the area so thoroughly that exposures of those rocks are hard to find.

Mining began in the Seattle area almost immediately after the first coal discovery in 1853. After a few years of intermittent operation, production increased steadily until it reached a peak of almost one-half million tons yearly in 1907. Then competition from cheap California oil drove production from the Seattle area coal fields steadily downward for decades.

Now that oil is no longer cheap, the Puget Sound coal fields once again have a future. Unfortunately, the quality of most western Washington coal is poor. Furthermore, the rocks of the Puget group were crumpled into tight folds as the North Cascade micro-continent jammed into the west coast of North America. Interesting as those folds are, they do complicate mining. Most of the folded coal seams must be mined underground, because they are too steeply tilted and too discontinous for an open pit operation. Most future production of Puget Sound coal, like most in the past, will probably serve the local market.

Cle Elum, near the former east coast of the North Cascade subcontinent, has been a coal mining town since it was founded, about 1885. At first, the coal went into railroad locomotives, now it generates electrical power. Most of the coal comes from the Roslyn formation, which is probably the same age as the Puget group near Seattle. The coal at Cle Elum likewise appears to have formed in tidewater swamps along the coastal plain of the micro-continent. Then the docking micro-continent jammed these rocks into folds like those in the Puget Sound area, complicating mining in the same way.

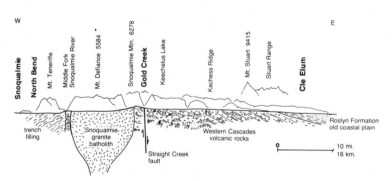

Section across the North Cascades along the line of Interstate 90.

Crossing the Cascades

Except for the sedimentary formations with their coal, litttle older bedrock that formed before the North Cascade micro-continent joined

North America is exposed along this route between Seattle and Ellensburg. Along most of the route through the higher part of the Cascades, the highway passes through volcanic rocks erupted from the Western Cascades, and intrusive granitic rocks that crystallized beneath those volcanoes. And there are a few younger volcanic rocks erupted from the modern High Cascades. As usual in large volcanic piles, sandstone and other sedimentary rocks exist interlayered with the volcanic material.

Cascade Volcanic Rocks

The Naches formation appears along the road from the area several miles west of Snoqualmie Pass all the way to Kachess Lake. It is a thick pile of volcanic rocks interlayered with beds of sandstone that accumulated between 40 and 45 million years ago, during Eocene time. Good exposures exist on both sides of Snoqualmie Pass, and along the east side of Keechelus Lake. The much younger granite of the Snoqualmie batholith intrudes the Naches formation.

Interstate 90 crosses the line of the Straight Creek fault at Kachess Lake, about halfway between Snoquaimie Pass and Cle Elum. Although the fault is one of the major structures in the North Cascades, its effects are hardly noticeable in this part of the range, where most of the rocks erupted from volcanoes long after the Straight Creek fault stopped moving, and buried most evidence of its existence.

Bedrock near the road along most of the route between Kachess Lake and Cle Elum is the Teanaway formation. It is yet another pile of volcanic rock erupted during Eocene time, and it also includes some sandstones. Basalt is the most abundant rock in the Teanaway formation, so most of the outcrops are very dark. For many years, rockhounds have come to the area between Cle Elum and Ellensburg to search the stream gravels for the blue agates that erode out of the Teanaway formation. Volcanic rocks generally contain gas bubbles that eventually fill with secondary minerals deposited from circulating groundwater, in some cases with agate. So the agates are not volcanic in origin, even though they form within volcanic rock. People find agates in stream beds because they are a variety of quartz, which resists both weathering and stream abrasion more successfully than the volcanic rocks in which they form, and so survive to become pebbles.

The high Mount Stuart Range, eroded in granite of the Mount Stuart batholith, retains its snow cover into mid-summer. This view looks north from Interstate 90 between Cle Elum and Ellensburg.

Plateau Basalt

About 7 miles east of Cle Elum, the highway passes through a narrow valley, where the Yakima River cut its canyon through basalt lava flows of the Columbia Plateau. That is one of the westernmost areas of plateau basalt, where the flows lapped into the lower parts of old stream valleys in the eastern margin of the North Cascade subcontinent. From that area to Ellensburg, the road crosses the western Columbia Plateau.

Below Thorp Prairie, the Yakima River breaks out into the broad Kittitas Valley, and follows it into Ellensburg. The Kittitas Valley is the trough of a broad synclinal down-fold in the Miocene basalt lava flows of the western margin of the Columbia Plateau. The corresponding anticlinal arches are the Mission and Manastash ridges in the distance north and south of the highway. Miocene basalt lava flows of the Columbia Plateau form the bedrock in both of those ridges.

Deposits of Pliocene gravel several million years old, and of much younger glacial sediments, completely bury the bedrock in the floor of the Kittitas Valley. The Pliocene gravel is a souvenir of the time when the climate was extremely arid throughout the Pacific Northwest. Folding of the basalt tilted the Pliocene gravel on the slopes of Manastash and Mission ridges.

144

U.S. 12
Chehalis—Yakima
145 mi. 230 km.

This route starts in the Willapa Hills, and then crosses the Cascades to the western edge of the Columbia Plateau. A few exposures of the older rocks of the North Cascade subcontinent peek out from under the deep cover of younger volcanic rocks.

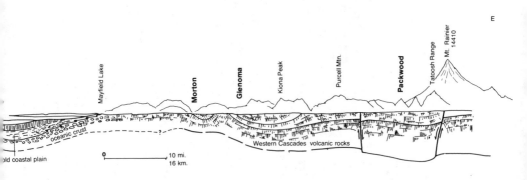

Section along the line of U.S. 12 from its junction with Interstate 5 to the edge of Mount Rainier National Park.

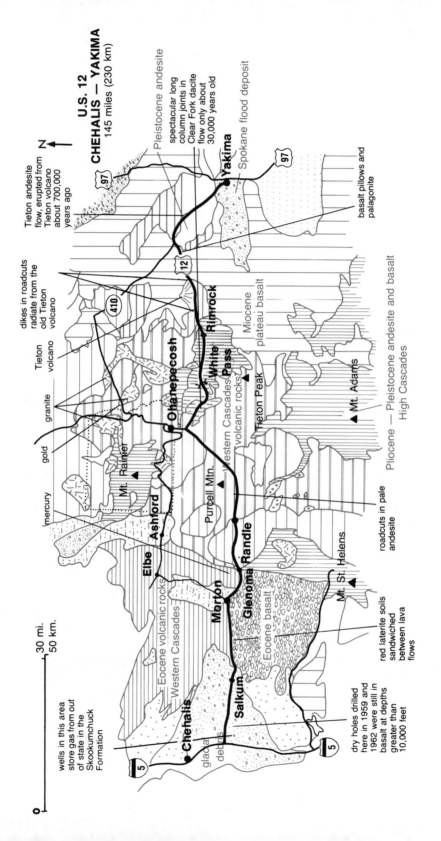

U.S. 12
CHEHALIS — YAKIMA
145 miles (230 km)

N

12

Pleistocene andesite

Tieton andesite flow, erupted from Tieton volcano about 700,000 years ago

spectacular long column joints in Clear Fork dacite flow only about 30,000 years old

97

Yakima

Spokane flood deposit

97

basalt pillows and palagonite

dikes in roadcuts radiate from the old Tieton volcano

Tieton volcano

granite

gold

410

Ohanepecosh

Rimrock

White Pass

Miocene plateau basalt

mercury

Mt. Rainier

Western Cascades volcanic rocks

Tieton Peak

▲ Mt. Adams

Pliocene — Pleistocene andesite and basalt High Cascades

Elbe

Ashford

Purcell Mtn.

roadcuts in pale andesite

30 mi.
50 km.

Morton

Randle

Glenoma

Eocene basalt

▲ Mt. St. Helens

wells in this area store gas from out of state in the Skookumchuck Formation

Eocene volcanic rocks

Western Cascades

Chehalis

Salkum

glacial debris

red laterite soils sandwiched between lava flows

5

5

dry holes drilled here in 1959 and 1962 were still in basalt at depths greater than 10,000 feet

0

The Willapa Hills

Between Interstate 5 and the vicinity of Glenoma, U.S. 12 crosses the eastern part of the Willapa Hills, where the bedrock consists entirely of material that formerly lay on the floor of the Pacific Ocean.

However, none of the old ocean floor appears at the surface between the interstate highway and Salkum. That western end of the route crosses a segment of the oceanic crust that dropped along faults to form a broad basin. Then meltwater from ice age glaciers dumped deep deposits of sand and gravel in the floor of the basin.

Rocks of the old ocean floor appear along the road between Salkum and Glenoma. Most are rather greenish pillow basalts that erupted far offshore in the crest of the oceanic ridge. There are also a few exposures of the dark gray muddy sandstones that accumulated on top of the pillow basalt. Fossils in the sandstone are the remains of animals known to have lived during Eocene time, perhaps about 50 million years ago. That was about the time the North Cascade micro-continent, which adjoins this expanse of oceanic crust, became part of North America. Evidently, this area was still on the floor of the ocean then.

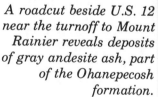

A roadcut beside U.S. 12 near the turnoff to Mount Rainier reveals deposits of gray andesite ash, part of the Ohanepecosh formation.

The Western Cascades

Near Glenoma, the old oceanic crust disappears beneath Western

Cascades volcanic rocks, which form most of the bedrock almost to White Pass. The volcanic rocks along this part of the route are dark andesites, mostly lava flows, which break into distinctive vertical columns, and mudflow deposits, which are simply masses of rubble hardened into rock. They belong to the Ohanepecosh formation, and erupted during Oligocene and early Miocene time, between about 35 and 15 million years ago. Although the bedrock is volcanic, the hills and valleys owe their form to stream and glacial erosion.

Younger Volcanic Rocks

Spindly columns of the Clear Fork flow, which poured down the valley of the Clear Fork about 30,000 years ago, exposed at a rest area beside U.S. 12 just east of Mount Rainier National Park.

More recent eruptions of the High Cascades built Mount Rainier, which occasionally appears high on the skyline north of the highway. Spiral Butte, north of the road about 5 miles west of Rimrock Lake, is one of the few High Cascades volcanoes near the highway. It is a complex of andesite lava flows and domes, actually a small volcano that must have erupted sometime during the ice ages, and blocked the drainage to impound Dog Lake. Spiral Butte andesites appear in roadcuts for a mile west and about two miles east of White Pass.

The North Cascade Subcontinent

In the area about 2 miles west of White Pass, and again all along Rimrock Lake, U.S. 12 passes through rocks that clearly belong to the North Cascade subcontinent. Geologists call them the Russel Ranch formation. Similar rocks probably lie beneath much of the Cascade volcanic pile, but this is the only area in this part of the range where they appear at the surface. Exposures are poor, but there are few roadcuts in dark gray sandstone of the kind that accumulates on the ocean floor, and of greenish pillow basalts, clearly old oceanic crust. Sandstone is most abundant near the western end of Rimrock Lake, and in the area west of the pass. Nice roadcuts in the greenish pillow basalts exist near the east end of Rimrock Lake.

The oceanic character of the Russel Ranch formation, and its severely deformed condition, suggest that it was stuffed into the trench that existed along the west coast of the North Cascade micro-continent. The age of the rocks is unknown, but they were almost certainly in the trench long before the micro-continent joined North America. The Russel Ranch formation probably correlates with similar rocks exposed west of the Straight Creek fault along Interstate 90 and U.S. 2.

The Fifes Peak and Tieton Volcanoes

Through a distance of about 5 miles east of Rimrock Lake, the highway passes through the Tieton basin, where glacial deposits and landslide debris cover the bedrock. For about 6 miles east of the Tieton basin, the road passes through rocks erupted from the Fifes Peak volcano, a part of the Western Cascades. Virtually every remnant of the volcanic peak has been eroded off the landscape, but enough of the rocks remain to document its former existence. Nevertheless, part of the Tieton volcano, which has been extinct for a least 25 million years, does survive in the hills north of the highway.

The Columbia Plateau

Somewhere approximately 10 miles west of Naches, U.S. 12 leaves the Cascades and enters the Columbia Plateau, which it crosses the rest of the way to Yakima. Flood basalts of the Columbia Plateau lapped onto the older volcanic rocks of the Western Cascades, and then younger volcanic rocks of the High Cascades covered parts of the plateau basalt. So the rocks interfinger, and the province boundary makes a ragged line on the map. High ridges north and south of the highway are anticlinal arches folded into the plateau basalts, and still rising. Between Naches and Yakima, the road follows the valley of the Naches River down the trough between the two arches.

Prominent cliffs with the conspicuous vertical colonnades typical of lava flows rise southwest of the road along much of the route between Naches and Yakima are the Tieton Canyon flows. There are two, erupted in the Goat Rocks area near Mount Rainier about 700,000 years ago. They poured east almost 50 miles down an old river valley, filling it with andesite. Since then, erosion has stripped away the old valley wall, leaving its filling of resistant andesite standing as a ridge.

Rocks in the hills north of the road between Naches and Yakima belong to a complex assortment of volcanic debris and gravelly sediments that geologists call the Ellensburg formation. It appears to be the remains of large alluvial fans that formed during the dry period of late Miocene and Pliocene time, between about 10 million and 3 million years ago, when gravels were deposited in every part of the region.

Washington 410
Tacoma — Mount Rainier — Yakima
67 mi. 108 km. 69 mi. 110 km.

Tacoma, Puyallup, and Enumclaw are in the Puget Sound lowland, in an area filled with deep deposits of glacial debris. Those sediments are at least 2000 feet deep in the Tacoma area, but thin to a few hundred feet near Enumclaw. East of Enumclaw, the highway crosses a large area of Western Cascade volcanic rocks.

The Osceola Mudflow

Buckley and Enumclaw stand on the Osceola mudflow, and the road follows it along the floor of the White River valley between Enumclaw and the Greenwater area. It is among the largest mudflows geologists have found anywhere in the world. Geologists have traced literally dozens of lesser, but still very large, mudflows in the valleys that drain Mount Rainier. Of all volcanic hazards, mudflows probably pose the greatest threat to life and property in the Puget Sound lowland.

151

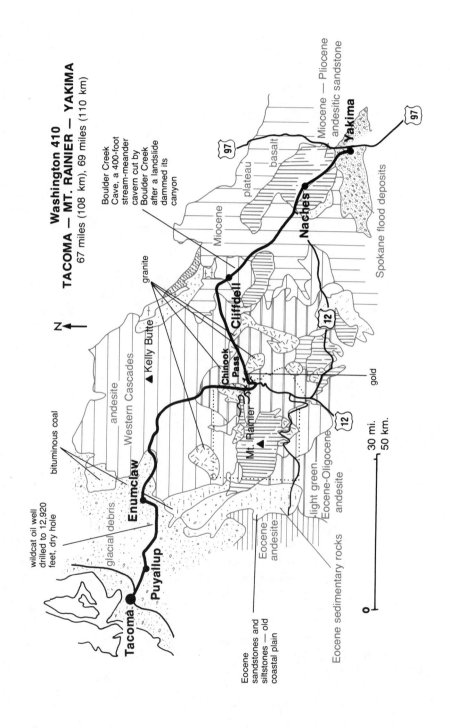

Washington 410
TACOMA — MT. RAINIER — YAKIMA
67 miles (108 km), 69 miles (110 km)

N

Boulder Creek Cave, a 400-foot stream-meander cavern cut by Boulder Creek after a landslide dammed its canyon

granite

Kelly Butte

andesite

Western Cascades

bituminous coal

wildcat oil well drilled to 12,920 feet, dry hole

glacial debris

Tacoma

Puyallup

Enumclaw

Chinook Pass

Cliffdell

Mt. Rainier

Naches

Yakima

97

97

12

12

gold

Miocene

plateau

basalt

Miocene — Pliocene andesitic sandstone

Spokane flood deposits

light green Eocene-Oligocene andesite

Eocene andesite

Eocene sedimentary rocks

Eocene sandstones and siltstones — old coastal plain

30 mi.
50 km.

0

Mudflow deposits exposed in roadcuts look very much like glacial till. They consist of a disorderly mixture of large and small rock fragments mixed together without internal layering. Like glacial moraines, the surfaces of mudflows are typically littered with transported boulders. Geologists distinguish between moraines and mudflows by plotting their distribution on a map to reconstruct the events that created them.

Glaciation

Most of the route passes through valleys that contained large glaciers during the last ice age, and clearly show the effects in their deeply gouged forms. One glacier filled the White River valley from Mount Rainier all the way to the Puget Sound lowlands. Another poured east down the American River to the area about 3 miles west of the turnoff to Bumping Lake where the road passes through roadcuts in rubbly glacial till.

The Western Cascades

Virtually the entire route from Enumclaw to U.S. 12 crosses old volcanic rocks erupted while the Western Cascades were active, and small areas of granite of about the same age. Most of the rocks between Enumclaw and Chinook Pass in Mount Rainier Park are dark andesites of the Ohanepecosh formation, typically mixtures of ash and rock fragments that formed as mudflow deposits. Geologists call those unsightly rubbles agglomerates.

An exposure of agglomerate.

Between Chinook Pass and the mouth of the American River, just above Cliffdell, the road passes pale volcanic rocks that belong to Fifes Peak and Stevens Ridge formations, also part of the Western Cascades. Like the darker andesites, most of these pale rhyolitic rocks are rubbles of rock fragments and ash, old mudflows.

Granite

Between American River and Chinook Pass, the road crosses several granite plutons. The rock is uniform pale gray at a distance, white peppered with black crystals of biotite and hornblende at close range. In one area about 4 miles east of Chinook Pass, pale gray granite contains numerous black inclusions of andesite several inches in average diameter. They are chunks of older volcanic rock that got caught up in the rising magma. All the granite is about the same age as the nearby volcanic rocks, and formed during the same period of volcanic activity. Regard the granite as rhyolite that did not erupt, or think of the rhyolite as granite that did erupt.

Plateau Basalt

Washington 410 crosses two areas of basalt that belongs to the Columbia Plateau, one small expanse between American River and Cliffdell, another much larger one in the easternmost 14 miles of the route. Recognize the lava flows by their dark solidity and their ten-

Wavy basalt columns stand in a roadcut in a Columbia Plateau lava flow near Naches. Gravel at the base of the roadcut is an old stream bed buried under basalt.

dency to break into conspicuous vertical columns. A rusty wash of iron oxide thinly stains the black basalt, giving many surfaces a dark brown color.

This is the western edge of the Columbia Plateau, where the basalt flows are crumpled into folds. Cleman Mountain, north of the highway, is an extremely tight anticlinal arch in parts of which the lava flows stand almost vertically. For about 15 miles west of Yakima, the highway approximately follows the axis of the synclinal trough between Cleman Mountain and the next anticlinal arch to the south.

Boulder Cave, just north of Cliffdell, formed as Devil Creek eroded down through a basalt lava flow, and into a soft deposit of gravel and soil buried beneath it. Then the creek scoured out the soft material, undercutting the lava flow. Finally, part of the undercut flow collapsed, partially damming the creek, and isolating the cave behind the deposit of rubble.

A Giant Landslide

Between 3 and 6 miles west of the junction between Washington 410 and U.S. 12, the highway passes a giant landslide. Watch for hummocky terrain along the north side of the road. The slide moved down the south slope of Cleman Mountain where the tilted layers of basalt lie nearly parallel to the slope of the mountain, a situation that fosters landsliding.

Washington 706, 123
Elbe — Mount Rainier Park — U.S. 12
56 mi. 90 km.

The drive between Elbe and U.S. 12 passes from the old coastal plain of the North Cascade micro-continent through the Western Cascades, and the southern edge of Mount Rainier National Park. Rocks along the road provide interesting glimpses into several aspects of Washington geology. The high country contains outstanding glacial scenery. See maps for U.S 12, Chehalis to Yakima and Mount Rainier National Park.

The Old Coastal Plain

Between the area about 5 miles east of Ashford and Elbe, the route crosses folded layers of sandstone, mudstone, and shale, that belong to the Puget group. These rocks accumulated as deposits of fine sand, mud, and clay along the western coastal plain of the North Cascade micro-continent. They date from Eocene time, between about 50 and 40 million years ago, when the micro-continent was closing on North America. Here, as elsewhere, the Puget group includes beds of coal and volcanic rocks, evidence that the micro-continent had swamps along its coast, and erupting volcanoes along its crest.

A small quarry exposure of flaky black shale in the Puget group.

Folds in the Puget group formations between the park and Elbe all trend generally northwest. Evidently, a force applied from the southwest crumpled the old coastal plain northeastward against the solid buttress of continental crust. That folding probably happened as the micro-continent jammed into North America, and the direction of folds suggests that it came in from the southwest.

The Western Cascades

Volcanic rocks of the Western Cascades lap over the older Puget group formations in the area just west of Mount Rainier Park. Most of those along the road belong to two major rock units: the Ohanepecosh formation, which consists mostly of dark andesites, and the Stevens Ridge formation, which contains a large proportion of light andesite and rhyolite. Distinguish them by their dominant colors. Most of the exposures of Stevens Ridge formation are along about 8 miles of road south of Stevens canyon.

This pale green volcanic breccia, or agglomerate, consists of dark chunks of andesite suspended in a matrix of volcanic ash. It is a mudflow deposit, part of the Ohanepecosh formation, exposed in Mount Rainier National Park.

A thick layer of red soil, laterite of the kind that forms in the wet tropics, separates the Ohanepecosh and Stevens Ridge formations. It is exposed in a large roadcut two miles north of the bridge across Nickel Creek. Evidently, there was quite a long period of weathering and erosion between eruption of the dark andesites and the lighter colored rocks.

The Tatoosh Pluton

Pinnacle Peak in the Tatoosh Range as it appears from near Paradise, on the south flank of Mount Rainier.

Although the highway enters and leaves Mount Rainier National Park on old volcanic rocks of the Western Cascades, much of its path through the southern part of the park crosses the Tatoosh pluton. It is granite, and very easy to recognize. Look for exposures of pale gray rock, mostly in the general area between the Nisqually River and Sunbeam Falls. Like all granite, it consists mostly of white feldspar, which is liberally salted with black needles of hornblende and flakes of biotite. And there are some grains of quartz that look like gray glass. Age dates show that the Tattosh pluton is about 17 million years old, late in the period of the Western Cascades.

Erosion stripped the Tatoosh pluton of its original cover of Ohanepecosh formation volcanic rocks during the long intermission between the end of volcanic activity in the Western Cascades, and its resumption in the modern High Cascades. Now Mount Rainier perches on that old erosion surface and in this area of the park we see a bit of the Tatoosh pluton peeking out from beneath the big volcano.

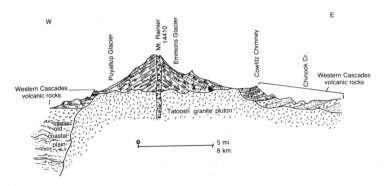

Section through Mount Rainier.

High Cascade Rocks

Of course, all of Mount Rainier consists of young Volcanic rocks erupted during the last million or so years. The road crosses those rocks in the Ricksecker Point area of the park, and also just east of the Reflection Lakes. They are dark andesites.

Rampart Ridge, which rises west of the highway along the route through the Nisqually River Valley north and south of Longmire, is one of the oldest parts of Mount Rainier. It consists of andesite lava flows that filled deep valleys eroded in the Ohanepecosh formation during the eruptions that first began to build Mount Rainier. Only after those valleys were filled could the volcano begin to rise above the rugged landscape on which it stands. The solid andesite of Rampart Ridge resists erosion better than the Ohanepecosh formation, which consists largely of rubbly mudflow deposits. So the solid andesite stands up in relief as erosion preferentially attacks the less resistant rocks. A valley of a million or so years ago is now a high ridge.

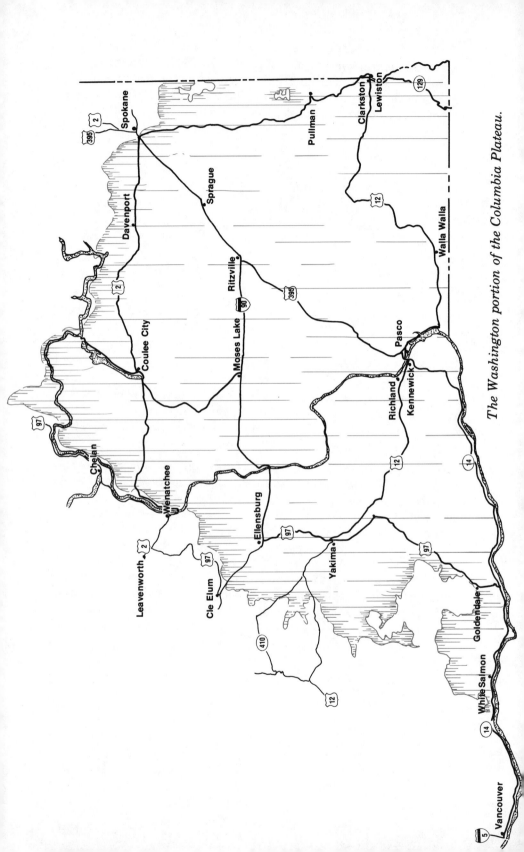

The Washington portion of the Columbia Plateau.

V
THE COLUMBIA PLATEAU

Most of central and southeastern Washington lies in the northern part of the Columbia Plateau, one of the world's largest and most spectacular volcanic provinces. It extends into the western fringe of Idaho, through most of the eastern half of Oregon, and into the northeastern corner of California. The Washington part of the Columbia Plateau lies between the Cascades and the Rocky Mountains, and south of the Kettle and Okanogan highlands.

Although we call it a "plateau," the area is actually a topographic basin enclosed within mountains. However, the landscape within that basin consists largely of hills with flat tops that do form many small plateaus. In any case, the expression "Columbia Plateau" is so deeply embedded in common use that any other would seem even more misleading.

The Giant Volcano

No one has clearly explained why the Cascades snuffed out while the volcanoes of the Columbia Plateau were active, and then revived when activity ceased in the Columbia Plateau. Neither does anyone know why the Snake River plain in Idaho, an eastern extension of the Columbia Plateau, remains active. Whatever happened, there is no doubt that the largest volcano of the Columbia Plateau flooded much of central and eastern Washington with basalt lava flows erupted from a center in southeastern Washington and nearby Oregon. We will call it the Grande Ronde volcano because the basalt dikes that fed its lava flows are best exposed in the canyon of the Grande Ronde River, on the border between Washington and Oregon.

Lava erupted from the Grande Ronde volcano poured down its flanks to flood east into the mountain valleys of western Idaho, north to the mountains of northeastern Washington, and west to the edge of the North Cascade subcontinent. Some flows followed the old valley of the Columbia River, now beneath Mount Hood, all the way to the Pacific. Many lava flows ponded in eastern and central Washington to form great pools of molten basalt a hundred or more feet deep that covered thousands of square miles.

There were many lava flows, hundreds of them, but they need not have been frequent. One large lava flow every few thousand years would have been more than enough to build the Columbia Plateau during the 4 million or so years of its most intense activity. Indeed, we can see quite easily that many of the flows were widely separated in time simply by looking at the soils sandwiched between them.

Floods of Basalt

A typical roadcut in a basalt lava flow. The columns range from the size of large fenceposts to that of barrels.

All basalt is black, fine-grained, and hard. Nevertheless, there are varieties of basalt, and geologists who work on the Columbia Plateau develop a sensitive eye for the subtle differences that distinguish one lava flow from another. They follow individual lava flows in the field, from one exposure to the next. Through years of such painstaking work, geologists have traced the full extents of many individual flows. All are big, some are truly enormous.

Many individual lava flows contain dozens of cubic miles of basalt thinly spread over hundreds, even thousands of square miles. Geologists call such overwhelming lava flows flood basalts. No basalt eruptions remotely comparable in volume have happened anywhere in the world during the period of recorded human history. Even without eyewitness accounts of large flood basalt eruptions, we can imagine them fairly accurately. The eruptions probably began as a crack several miles long and several yards wide opened in the earth's surface. Then a row of cinder cones developed along the fissure as red hot basalt lava welled up and spread across the landscape. We can see basalt filled fissures in areas of the Columbia Plateau where the lava emerged.

Dike Swarms

Geologists call a fracture filled with igneous rock a dike. Thousands of basalt dikes exist in southeastern Washington and nearby Oregon. They contain basalt identical to that in the flows in the Washington part of the Columbia Plateau, so it seems reasonable to conclude that the dikes fill the fissures that erupted those flows. The average width of those dikes is several yards, and they all follow a generally north to south trend. Dikes, basalt, and the Columbia Plateau all exist because something was stretching the earth's crust in an east to west direction during late Miocene time.

Origin of the Basalt Magma

Remember that the earth's crust is a thin rind of rock on the much denser peridotite rock in the mantle. It is so hot within the mantle that the rocks there would melt, were they not under great pressure. If the earth's crust stretches, it also thins

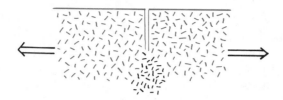

Stretching the earth's crust opens fissures that relieve pressure on the hot rocks at depth, thus permitting them to melt.

and cracks, and that reduces the pressure on the extremely hot rocks at depth. Then those rocks start to melt, producing molten basalt magma that rises because it is lighter than the solid rocks that surround it. It seems strange at first thought that partial melting of peridotite should produce basalt, a different rock. However, the phenomenon is familiar. Everyone has sucked the juice, a partial melt, out of a popsicle, leaving white ice on the stick. Similarly, the juice that rises from partially melted peridotite in the mantle is basalt magma.

So the long fissures that basalt magma wells out of when erupting may themselves cause the magma to melt at depth. Therefore, whatever force stretched the earth's crust must cause both the fissure and the magma. If it were possible to count all the dikes and add their thicknesses together, the sum would equal the amount of crustal stretching.

What Lies Beneath All the Basalt?

Few things could be more opaque than the thick stack of volcanic rocks that comprise the Columbia Plateau. They make it extremely difficult to say much about what lies hidden beneath them, what existed in the region before those rocks formed. Nevertheless, the question is important, especially to people who suspect that the rocks at depth may contain reserves of oil and gas.

No doubt, the western edge of the North American continent lies beneath the lava flows along the eastern edge of Washington. Numerous high hills composed of old continental rocks rise like islands above the lava flows. Because the

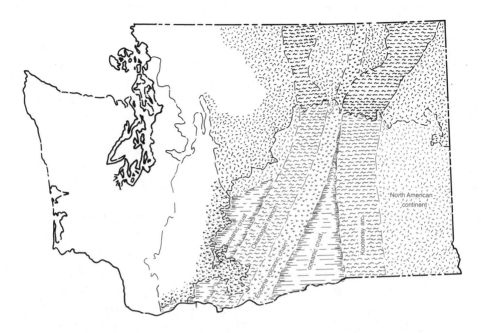

A somewhat conjectural map showing the rocks that might appear if we could peel all the volcanic rocks off the Columbia Plateau.

Kootenay arc formed as an old continental shelf squashed against the former western margin of the North American continent, there is every reason to suspect that it likewise continues south through Washington, although there are no exposures.

The Okanogan micro-continent, which joined North America sometime around 100 million years ago, probably extends south at least some distance beneath the Columbia Plateau. But there is no reason to assume that it is as continuous as the Kootenay arc. It seems far more likely that the Okanogan subcontinent ends somewhere beneath the northern part of the Columbia Plateau. If so, an expanse of oceanic crust connected to the Okanogan subcontinent probably extends southward beneath much of the eastern part of the Columbia Plateau. Like most oceanic crust, that probably has on it a thick accumulation of sedimentary rocks that could contain petroleum. A similar expanse of oceanic crust extends south from the North Cascade subcontinent through much of western

Oregon, and probably also beneath a small western slice of the Columbia Plateau in Washington.

After the Okanogan subcontinent landed, the oceanic trench shifted west to a line that passed through the Okanogan Valley, and certainly continued south through Washington and Oregon, and down the coast of California. Rocks stuffed into that trench probably exist beneath the western edge of the Columbia Plateau. Remnants of the volcanic chain that formed east of that trench survive in the Eocene volcanic rocks of the Republic graben north of the Columbia Plateau, and in the Ochoco Mountains of central Oregon. Similar Eocene volcanic rocks probably exist beneath the western part of the Columbia Plateau along a broad zone connecting the Ochoco Mountains with the Republic graben. Two deep wildcat wells drilled near Yakima apparently penetrated that old volcanic chain.

Eastern Washington may well have been a bay of the ocean 50 million years ago when those Eocene volcanoes were erupting along a line connecting the Republic graben with the Ochoco Mountains. If so, it would have trapped a thick accumulation of sedimentary rocks, which may well exist beneath the eastern part of the Columbia Plateau. If there was a bay in eastern Washington during Eocene time, that area probably has good potential for producing oil.

The Western Cascades included large volcanoes that erupted enormous volumes of rhyolite ash. Clouds of white volcanic ash drifted eastward on the wind and settled to make thick deposits. That ash is exposed in large areas of central Oregon, where geologists call it the John Day formation. Although no exposures of the John Day formation exist in Washington, it probably lies buried beneath the younger volcanic rocks.

So a deep well drilled somewhere in the central part of the Columbia Plateau would probably penetrate at least several thousand feet of lava flows. Then it would pass into the John Day formation, which could be as much as several thousand feet thick. Beneath the John Day formation, it may well enter sedimentary rocks that accumulated in sea water, and therefore could contain oil or gas. As we write this, no such well has been drilled. However, such drilling will almost certainly happen before too many years pass, and then everyone can find out whether this analysis is correct.

Laterite Soils, Porcelain Clay, and the Clarity of Beer

Nearly everywhere on the Columbia Plateau, conspicuous red streaks appear on canyon walls and in roadcuts. Those are old soils developed on one lava flow, and buried beneath the next. In a few areas of the Willapa Hills in western Washington, the red soils were never buried, and remain at the surface. It is also commonplace to see layers of white sediment, old lake beds, sandwiched between lava flows. Together, the red soils and white lake beds record an interval of warm and wet climate during late Miocene time, when the Columbia Plateau was active.

The red soils are laterite, a type of soil that consists mostly of red iron oxide, aluminum oxide, and a clay mineral called kaolinite. Laterites form only in regions with warm and very wet climates, nowhere in regions as cool or as dry as the modern Columbia Plateau. Near Astoria, where the Miocene flood basalts extend west of the Cascades, the soil locally becomes bauxite, an extreme form of laterite so rich in aluminum oxides that it is aluminum ore. Laterite soils form today in much of the southeastern United States, and Central and South America, bauxites only in the tropics.

Evidently, the Columbia Plateau drained west during late Miocene time, as it still does. The big lava flows blocked streams to impound large lakes and swamps that accumulated sediment. Geologists call all the lake deposits sandwiched between lava flows the Latah formation, after a small town south of Spokane where they were first carefully studied and described. The formation contains valuable industrial minerals.

Many deposits of the Latah formation consist largely of kaolinite clay eroded from the laterite soils. Red iron oxide that also eroded from those soils dissolved in the slightly acid lake and swamp water, leaving the pure white kaolinite. It is raw material for porcelain, fire brick, and other ceramic products. Some of the lakes filled with deposits of sparkling white diatomite, a sediment composed of the microscopic skeletons of floating algae, which look under a high powered microscope like especially intricate lacework. Diatomite is valuable for making very fine filters, such as those used to strain the yeast out of beer. It also makes good fireproof insulation in safes.

Tropical Plants

Some exposures of the Latah formation contain beds of white volcanic rhyolite ash that preserve impressions of leaves as nicely as though they were pressed in a dictionary. Evidently the leaves were lying scattered on the bottom of a lake when an ash fall buried them. Many of those leaves are from species of trees that thrive today in the Caribbean region, and they are further evidence that Washington had a wet and tropical climate during late Miocene time. In a few places where lava flows buried lake beds full of water-soaked logs, we now find deposits of petrified wood that tell a similar story.

Imagine those palm fringed late Miocene lakes of eastern Washington set in a landscape covered with a lush forest of tropical hardwood trees. And now imagine a basalt lava flow searing across that landscape, clouds of steam hissing out of a boiling lake, enormous plumes of smoke rising from burning trees floating on the red hot lava. Those must have been among the largest fires ever set on this planet, and the drifting smoke must have smudged the air for weeks on end. Then, hundreds or thousands of years passed, new soil formed, and new forests grew, before the next flood of basalt created another swath of black countryside.

The Dry Time

In well watered regions, the streams have enough flow to carry their load of sediment all the way to the ocean, leaving very little behind to accumulate on the continent. The sparse plant cover of dry regions does not protect the ground from erosion, and the streams of those regions do not have enough flow to carry all the sediment they receive to the ocean. So deposits of sediment, including great quanitities of gravel, accumulate on the land where the climate is dry.

Throughout the Pacific Northwest, conspicuously on the Columbia Plateau, extensive deposits of gravel tell of a long period when the climate was dry. In many places, that gravel lies directly on the red laterite soils, making it clear that the dry time followed the episode of tropical climate. In some places, geologists have found in the gravel the fossil bones of animals such as horses and camels that thrive in dry regions. And the remains are those of animals known to have lived during Pliocene time, between 10 and 2 million years ago, in round numbers.

All the younger gravels in Washington, and elsewhere in the region, are clearly associated with ice age glaciation, and so do not tell of a dry climate. The evidence suggests that the dry climatic period ended with the onset of Pleistocene glaciation sometime between 2 and 3 million years ago.

The Columbia River

The Columbia River enters Washington along the eastern margin of the Okanogan subcontinent. Then it follows the edge of the Columbia Plateau for many miles, first tracing the southern margin of the Okanogan subcontinent, then following the eastern edge of the North Cascade subcontinent about as far south as Wenatchee. To think of it another way, the river skirts the flanks of the old Grande Ronde volcano. That is logical if we picture the volcano sloping down to the west and north from its summit near the southeastern corner of Washington. The line where its flanks lapped onto the bounding highlands would have been the lowest part of the region, therefore the path the drainage would follow.

The Folded Western Plateau

In the eastern half of the Columbia Plateau, the basalt lava flows still lie almost flat. Farther west, they buckle steeply into large folds, which trend in a generally east to west direction. Those folds appear in the landscape as high ridges: the Frenchman Hills, Saddle Mountains, Yakima Ridge, Rattlesnake Hills, all the long ridges that extend east from the Cascades into the Columbia Plateau. Each of those ridges is a fold, an anticline arched in the basalt lava flows during the last 10 million years as. northward movement of the west coast buckles the rocks. Such anticlinal arches more commonly appear as valleys than as ridges, because the fractured rocks in the crest of the arch erode very easily. Only in places where the folds are extremely young do they form ridges. These folds are so young that deposits of Pliocene gravel lie steeply tilted on their flanks. In fact, these folds are still growing.

South of Wenatchee, the Columbia River passes through several fold ridges in spectacular narrow gorges, as does the Yakima River between Ellensburg and Richland. Evidently, the rivers were in their present courses before folding raised the ridges, and maintained themselves in those courses by eroding the basalt as fast as the ridges rose across their paths.

Wind Blown Dust

The rich bounty of the Palouse loess. This overflowing elevator is at Creston, beside U.S. 2 west of Spokane.

A veneer of yellowish wind blown dust, geologists call it loess, covers large parts of eastern Washington. The dust is in old dunes, which survive as the rolling Palouse Hills of eastern Washington and westernmost Idaho. No soil is more fertile than that developed on loess, and few regions of North America produce crops to surpass those harvested in the Palouse Hills.

The wind blown dust of the Palouse Hills contains little evidence of its age except that it accumulated long after volcanic activity ceased in the Columbia Plateau, and sometime before the last ice age. The origin of the dust is equally debatable. Large clouds of dust blow off glacial outwash deposits during dry weather, and large loess deposits commonly form downwind from glaciated regions. However, dust storms also blow out of deserts, and loess deposits commonly form downwind from them too. It is easy to apply either explanation to the Palouse Hills. They lie immediately south of large glaciated regions, and immediately northeast of the extremely dry country in south central Washington and nearby Oregon.

The World's Greatest Floods

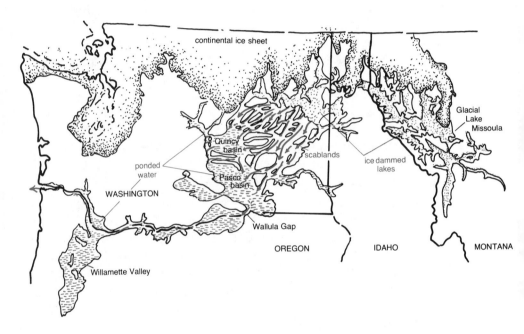

Glacial Lake Missoula, the glacier that dammed it, and the channeled scablands of eastern Washington.

Here we come to an episode so fantastic that it seems to perch on the thin edge between science and science fiction. J. Harlen Bretz, one of the giants of Washington geology, began to study the channeled scablands of eastern Washington during the 1920's. After much thought, he finally concluded that nothing but a catastrophic flood could have eroded those high channels. He called it the Spokane flood, and envisioned an overwhelming wall of water surging across eastern Washington. Most geologists were deeply skeptical for years, but Bretz and others have since shown beyond any reasonable doubt that the Spokane flood really did happen — about 40 times.

Glacial Lake Missoula

When Bretz first proposed his theory, he could not explain where the Spokane flood came from. Several years passed before another geologist found the source in a vanished lake in western Montana, Glacial Lake Missoula. The Clark Fork River flows northwestward through western Montana and the panhandle of Idaho into Pend Oreille Lake. As the glaciers of the last ice age approached their maximum size, an enormous glacier pushed south out of Canada into the Idaho Panhandle, and dammed the Clark Fork River at the site of Pend Oreille Lake. That probably happened sometime around 16,000 years ago. Water backed up into and flooded the mountain valleys of western Montana to create Glacial Lake Missoula, which rose to a maximum elevation of about 4350 feet. At that level the water behind the ice dam was 2000 feet deep, and the lake contained about 500 cubic miles of water.

Ice is terribly poor material for a major dam because it will float if the water behind it rises high enough. When the ice dam in northern Idaho floated and broke, it dumped the entire contents of Glacial Lake Missoula onto eastern Washington. The Spokane flood was the most catastrophic flood of known geologic record. Imagine a wall of water 2000 feet high with 500 cubic miles of water behind it heading with a thundering roar for Spokane and points southwest.

After the ice dam washed out, the glacier that had formed it continued to move south. Within a few years, the glacier dammed the river again to impound another Glacial Lake Missoula, which likewise suddenly drained as the ice dam floated and broke again. The valley walls above the Clark Fork River in western Montana bear numerous lake shoreline scars, perfectly horizontal benches daintily etched in the hillslopes. Those shoreline scars are too faint to count with confidence, but their number is between 25 and 40. Each one probably corresponds to a separate filling of the lake.

Each time the lake drained, it released another repetition of the Spokane flood. Those that corresponded to the higher shorelines were certainly larger than those that emptied lower levels of the lake. Geologists who have studied the flood-scoured valleys of western Montana estimate that the Clark Fork and Flathead rivers there carried between 8 and 10 cubic miles of water per hour while Glacial Lake Missoula emptied — all heading for the Columbia Plateau.

The Channeled Scablands

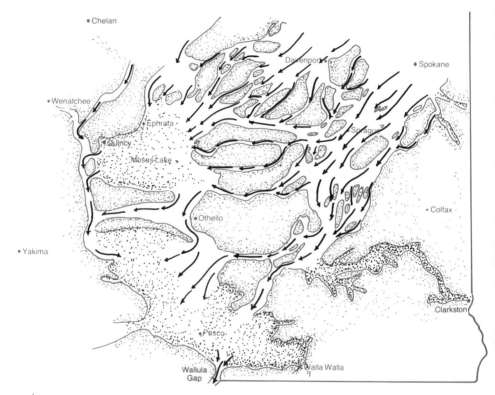

A broad view of the channeled scablands. The black arrows show flood-scoured channelways, the black dots show areas where flood waters backed up to form temporary slackwater lakes. The colored dots stipple areas that stood above the floodwaters.

The Spokane floods were so completely out of scale with the landscape of eastern Washington that they behaved as though a bucket of water had been dashed across a table top model.

173

Each great tide of water completely filled the normal stream valleys, and then washed across the divides in a headlong sweep across southeastern Washington to the Columbia River. In several places, the dry channels cross divides several hundred feet above the level of the modern drainage. Imagine those turbulent masses of muddy water shaking the ground as they thundered across the eastern part of the Columbia Plateau.

Those high channels contain stream rounded boulders of basalt as much as several feet in diameter, much too big for any of the modern streams in the area, indeed any ordinary stream, to move. The dry channels look so fresh that it seems as though water might have poured through them only last week. Bedrock exposed in them is almost unweathered. That freshness means that the scabland channels must be young, that they can not be remnants of some extremely ancient and long abandoned stream system.

Wherever the flood water spilled, it swept the yellow loess soil off the landscape, exposing the hard, black basalt beneath. In places, the flood swept around both sides of a loess hill, leaving it standing as a streamlined island in the black channeled scablands. Many of the loess islands, as well as the loess hills bordering the scablands, have benches on their slopes that record the high flood mark. The strong color contrast between the black basalt in the flood channels and the yellow loess islands is obvious from the air, even on satellite images.

The Spokane floods slowed in the broad valleys, such as the Quincy basin, and backed up in them to form temporary lakes that probably lasted for a few days with each flood. The floods also backed up all the streams that drain into the Columbia to form slackwater lakes in which thick deposits of sediment accumulated. The Snake River, for example, flooded to depths as great as 600 feet at Clarkston.

Like the roads that led to Rome, all the Spokane flood channels led eventually to the Pasco basin where the water backed up behind the bottleneck of Wallula Gap to form the biggest lake of them all. It reached depths of as much as 1000 feet. All that water drained through the Wallula Gap within a few days at a maximum rate of about 40 cubic miles per day. Another bottleneck at the Columbia Gorge also backed water up to a

ERTS satellite image of channeled scablands from an elevation of about 500 miles.

depth of almost 1000 feet during the day or so of maximum flood flow. The rush of water through the Columbia Gorge scrubbed most of the soil off the rocks there to an elevation of about 1000 feet, leaving the lower canyon walls almost bare rock. Rapid erosion during the flood cut the walls of the Columbia Gorge back far enough to amputate the lower ends of tributary stream channels, forcing them to enter the gorge over small waterfalls. Multnomah Falls, on the Oregon side of the river, is the most familiar example. The last major bottleneck was the sharp northward bend of the Columbia River at Vancouver and Portland, where the ice jammed to form another temporary dam. Water backed up to a depth of about 400 feet, and flooded the entire northern half of the Willamette Valley in Oregon.

Flood Deposits

As the Spokane floods stripped loess soil off the countryside in their path, and ripped up the basalt beneath to make the channeled scablands, they also dumped that material elsewhere to make equally spectacular flood deposits. Fine-grained sediment tended to accumulate in the slackwater lakes. In those sheltered places, the successive layers of sediment piled up without disturbing those below to make a complete record of the Spokane floods. In several places, geologists have counted as many as 40 flood deposits. That number corresponds closely to the count of shoreline benches on the mountainsides of western Montana.

The coarser basalt debris tended to accumulate in valleys, such as the Pasco and Quincy basins, that lay in the path of the flood. Some of the broader flood channelways also contain deposits of sediment, many of which express the incredible rush of water in trains of giant ripple marks on their surfaces. These look just like the ripple marks we see in the sand on the bottom of a stream, except that they measure hundreds of feet from crest to crest, and are as much as 30 feet high. Giant ripple marks are so big that it generally takes quite an investment of effort and imagination to see them on the ground. But it is easy to see them from the air or from a distance in the glancing light of early morning or late afternoon, and from above they do indeed look like giant ripple marks.

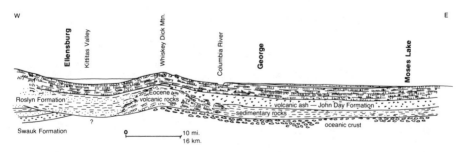

I-90
Ellensburg — Spokane
188 mi. 300 km.

All of the bedrock between Ellensburg and Spokane is basalt lava flows erupted from southeastern Washington and northeastern Oregon during late Miocene time, as volcanic activity in the Columbia Plateau was nearing its end. As everywhere, the basalt is black, fine-grained, and too hard to scratch easily with a knife. Its tendency to break into vertical columns makes cliff exposures distinctive, even at a distance.

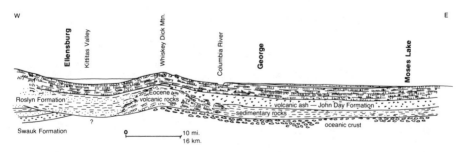

Section along the line of Interstate 90 between Ellensburg and Moses Lake.

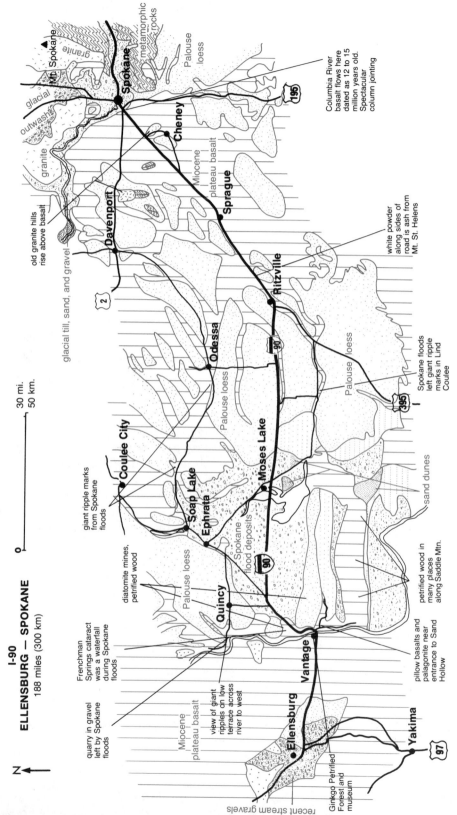

I-90
ELLENSBURG – SPOKANE
188 miles (300 km)

0 ___ 30 mi.
___ 50 km.

N →

Columbia River basalt flows here dated as 12 to 15 million years old. Spectacular column jointing

white powder along sides of road is ash from Mt. St. Helens

Spokane floods left giant ripple marks in Lind Coulee

old granite hills rise above basalt

glacial till, sand, and gravel

Mt. Spokane ▲

granite

metamorphic rocks

Palouse loess

glacial outwash

granite

Spokane

Cheney

Davenport

Miocene plateau basalt

Sprague

Ritzville

Odessa

Palouse loess

Palouse loess

Coulee City

Moses Lake

sand dunes

giant ripple marks from Spokane floods

Soap Lake

Ephrata

Spokane flood deposits

diatomite mines, petrified wood

Palouse loess

Spokane flood deposits

petrified wood in many places along Saddle Mtn.

Frenchman Springs cataract was a waterfall during Spokane floods

quarry in gravel left by Spokane floods

Miocene plateau basalt

view of giant ripples on low terrace across river from west

Quincy

Vantage

pillow basalts and palagomite near entrance to Sand Hollow

Ellensburg

Yakima

Ginkgo Petrified Forest and museum

recent stream gravels

Gingko Petrified Forest

The Gingko Petrified Forest with its museum and state park perches on a high bluff overlooking the Columbia River near Vantage, 30 miles east of Ellensburg. We rarely see much petrified wood associated with large basalt flows because trees generally float to the surface of the lava, and burn. That did not happen here because the lava covered waterlogged wood buried in the mud in the bottom of a shallow lake. After the flow buried them and their enclosing mud, those logs petrified, and we see them now remarkably well preserved.

Petrified wood consists mostly of silica in the form of the minerals chalcedony and opal. It forms because silica is extremely insoluble in acid solutions, but significantly soluble in alkaline solutions. The mud on the lake bottom contained volcanic ash, which made the water alkaline, and therefore able to dissolve silica. Meanwhile, the waterlogged wood buried in the mud became acidic as it started to decay. Therefore, any dissolved silica that diffused into the buried logs became insoluble, and precipitated as chalcedony or opal.

We often read that petrifaction involves a molecule for molecule replacement of wood by silica, but you can see by looking at the exhibits in the museum that no such thing could have happened. You could not replace all the bricks in a wall without changing its size and shape unless the new bricks were identical to the old ones. Silica molecules are not the same size and shape as those that constitute wood, so a molecule for molecule replacement of wood by silica would involve great distortion of the wood. Instead, we see exquisite preservation of delicate structures such as growth rings, even the original cell walls. In fact, much of the original wood still remains, and it is simply soaked with silica minerals, not replaced by them.

Somber cliffs eroded in basalt lava flows of the Columbia Plateau stand guard over the Columbia River north of Vantage.

No one knows how long it may have taken that waterlogged wood to petrify, but it need not have been very long. Some of the trees surrounding hot springs in Yellowstone Park are partially petrified right where they stand. Indeed, it seems likely that the logs at the Gingko Petrified Forest may have petrified rather quickly, because they show little sign of having been crushed beneath the burden of sediments and lava that buried them.

The Quincy Basin

Most of the route between George and Moses Lake crosses the Quincy basin, where the Columbia River and the Spokane floods deposited much of the debris they eroded from the Grand Coulee. No bedrock appears along this stretch of road because it lies buried beneath several hundred feet of sediment. Some of the floodwater drained south out of the Quincy basin through the Drumheller channels in the Potholes reservoir area south of Moses Lake. The rest drained west into the Columbia River over two enormous waterfalls: Potholes Coulee, 7 miles southwest of Quincy, and Frenchman Coulee, just west of I-90 six miles northeast of Vantage.

Potholes Coulee is the larger of the two, with a vertical drop of some 500 feet. Floodwaters poured out of the Quincy basin over the crest of Babcock Ridge, and down the slope to the Columbia River in a series of waterfalls and cascades. Frenchman Coulee was also a series of cataracts and waterfalls with a total vertical drop of about 500 feet down the side of Evergreen Ridge. It must have been quite a spectacle to see the water suddenly overflow from the Quincy basin, and roar down the slope toward the Columbia River. Deep holes below the vertical cliffs of the old waterfalls are the old plunge pools. Several small mines north of Frenchman Coulee produced diatomite from old lake deposits sandwiched between lava flows.

Giant current ripples about 10 feet high march down the surface of the Crescent Bar on the Columbia River north of Vantage. They formed as the Spokane floods passed.

The Big Ash Fall

The area between Moses Lake and Ritzville received the heaviest fall of volcanic ash, about 4 inches, after the May 18, 1980 eruption of Mount St. Helens. Evidently the force of the blast lofted most of the ash over the intervening countryside, and then dumped it into this part of east central Washington. On the morning after the big eruption, ash lay on this part of Washington like dirty gray snow that would never melt.

No one seems to have any use for volcanic ash, so there was nothing to do but get it out of the way. In town, that involved scraping it up, loading it onto trucks, and hauling it to a dump. Ritzville disposed of some 210,000 tons at a cost estimated at about $2.37 per ton. In rural areas, graders bladed the ash off the roads, and farmers plowed it into their fields. Some roads were resurfaced with a new topping of gravel.

When it fell, many people feared that volcanic ash might be dangerous. Volcanic ash is not poisonous, but one can easily imagine that the millions of microscopically sharp glass edges might damage delicate lung tissues. Anyone who could have gotten into central Washington on the morning after the ash fall with a big load of dustproof face masks could have made an instant fortune. We don't know whether the ash damaged any lungs, but there is no doubt that it did clog air filters and ruin bearings in vehicles and in farm equipment.

Scanning electron microscope photo of a particle of Mount St. Helens ash magnified 900 times. The holes are minute gas bubbles in volcanic glass, and the sharp edges between them the broken walls of gas bubbles. J.N. Moore photo

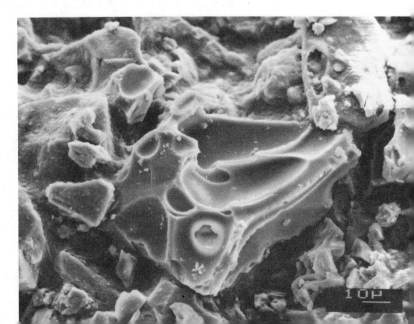

There was some hope that the ash might benefit the soil, and farmers in the area of the ash fall did get fine crops in 1980. However, there is little evidence that rhyolitic ash makes good fertilizer, and little real hope for significant long term benefit. Dark basaltic ash contains more of the nutrients plants need, and would have helped more.

Very few people expected the volcanic ash to act as an insecticide, but it did. Evidently, the sharp edges lacerate the insects' outer skin causing it to lose water and die. For weeks during the early summer of 1980, the entire area of the ash fall was remarkably free of flying and crawling insects, and of the birds that feed on them. Unfortunately, bees died too. The same sharp edges readily interlock once the ash is on the ground, making the deposit extremely stable, and explaining why it was still conspicuous on surfaces of roadcuts several years after it fell.

Scablands

Numerous good exposures of basalt appear along the road between Moses Lake and Spokane where it crosses the channeled scablands. The Spokane flood washed across this countryside as though a washtub were emptied over a child's sandbox. The effects are quite visible from the highway in the numerous flood channels scoured into basalt, dry now except for ponds and small lakes in the low places, in the ragged outcrops of basalt, and in the widespread absence of the Palouse loess. Sprague Lake, south of the road about 35 miles southwest of Spokane, fills one of the larger flood scoured basins.

Medical Lake, a small body of water north of the highway about 15 miles west of Spokane, fills another low spot in one of the scabland channels. The main outlet is through evaporation, so the water has become salty and slightly alkaline. Someone began bottling that water and selling it for medicinal purposes as early as 1885, thus giving the lake its name. We can find no evidence that the water or the salt extracted from it have any therapeutic value.

Between the dry channels, there are tracts of gently rounded Palouse Hills composed of wind blown loess that stood above the flood, and remain perfectly unscathed. The pattern is much more obvious from the air where one can see the branching and criss-crossing stream channels cutting harshly through the soft landscape of the loess hills. The black basalt exposed in the channels contrasts sharply with the tan and green of the hills.

U.S. 2
Wenatchee—Spokane
159 mi. 255 km.

U.S. 2 follows the east bank of the Columbia River about 20 miles north from Wenatchee, and then cuts straight across the Columbia Plateau to Spokane. Much of the route passes through or very near some of the most spectacular parts of the channeled scablands.

Except for a few knobs of granite in the Grand Coulee, all of the bedrock exposed along or near the road is basalt lava flows of the Columbia Plateau. As everywhere, the basalt columns are distinctive at a distance. At close range, the rock is fine grained and very dark gray or black — brownish black on weathered surfaces. All of these flows erupted from the Grande Ronde volcano in southeastern Washington and nearby Oregon.

Between Wenatchee and Coulee City, deep soils, and glacial sediments near Coulee City, cover most of the basalt and provide good agricultural soils. The road crosses the Grand Coulee at Coulee City, and then passes across rolling hills of wind blown soil, the Palouse loess, between Coulee City and Almira. Loess is excellent soil, and the fertile fields yield abundant wheat crops.

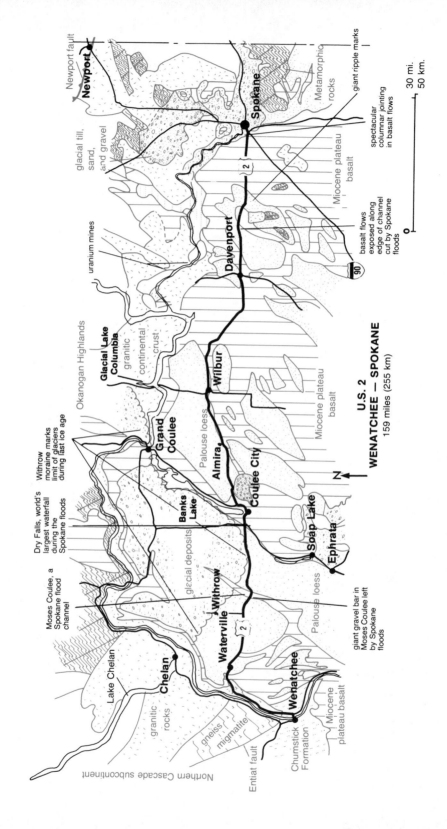

U.S. 2
WENATCHEE — SPOKANE
159 miles (255 km)

2

Newport fault

Newport

glacial till, sand, and gravel

Spokane

Metamorphic rocks

giant ripple marks

2

spectacular columnar jointing in basalt flows

Miocene plateau basalt

90

basalt flows exposed along edge of channel cut by Spokane floods

Davenport

uranium mines

Glacial Lake Columbia

granitic continental crust

Wilbur

Okanogan Highlands

Miocene plateau basalt

Palouse loess

Grand Coulee

Withrow moraine marks limit of glaciers during last ice age

Dry Falls, world's largest waterfall during the Spokane floods

Almira

Coulee City

N

Banks Lake

glacial deposits

Soap Lake

Ephrata

Moses Coulee, a Spokane flood channel

Lake Chelan

Withrow

Waterville

2

Palouse loess

giant gravel bar in Moses Coulee left by Spokane floods

Chelan

granitic rocks

gneiss

migmatite

Wenatchee

Miocene plateau basalt

Chumstick Formation

Entiat fault

Northern Cascade subcontinent

30 mi.
50 km.

0

Basalt pillow and palagonite complex exposed in a roadcut beside U.S. 2 a few miles east of Waterville. The dark blobs are basalt pillows and fragments of pillows that formed as the lava poured into a lake. The light material is yellowish palagonite, a soft rock full of clay that forms as steam reacts with basalt.

The higher hills visible in the distance south of Reardan contain metamorphic rocks similar to many in the area north of Spokane. They are old hills that stand above the surrounding flood of basalt, the westernmost exposures of the old North American continent.

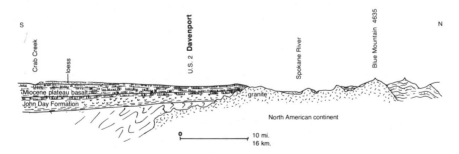

Section across the line of U.S. 2 at Davenport showing the flood basalts of the Columbia Plateau lapping onto the older rocks north of the highway.

185

RELICS OF THE ICE AGE

Moses Coulee

Moses Coulee, an abandoned channel big enough to hold a major river. Imagine it filled to the brim with the muddy torrents of the Spokane floods.

U.S. 2 crosses Moses Coulee in the area about 12 miles west of Coulee City. It is hard to miss, a broad and deep valley that looks like it should contain a major river, but no longer does.

There is no doubt that the Spokane floods helped erode Moses Coulee, because it contains sets of giant current ripples, and giant gravel bars. U.S. 2 crosses the north end of the most spectacular of those gravel bars in a great sweeping switchback on the west side of the coulee. The top of the giant gravel bar is about 300 feet above the floor of the coulee, so the floods that left it must have been at least that deep.

About 3 miles north of U.S. 2, at Jameson Lake, Moses Coulee disappears beneath the Withrow moraine, which marks the farthest advance of glaciers during the last ice age. The same area contains numerous drumlins, small hills made of glacial till and streamlined beneath the moving ice. Obviously, the coulee and its giant gravel bars are older than the moraine. So they must have formed sometime before the maximum of the last ice age, sometime before about 15,000 years ago. No one knows when Moses Coulee formed, but it is clearly older than Grand Coulee.

Grand Coulee

Sediments deposited in Glacial Lake Columbia. The alternating thin light and dark layers are varves that accumulated during the summers and winters of the last ice age, and record the passage of each year the lake existed. The thick beds are debris dumped into Glacial Lake Columbia from the Spokane floods.

The Grand Coulee, north and south of Coulee City, is one of Washington's geologic spectaculars. During the last ice age, a glacier blocked the Columbia River about 50 miles north of this area to impound Glacial Lake Columbia, which drained through the Grand Coulee. The Columbia River was much bigger then than it is now because the ice ages were wet, as well as cold. So the diverted river eroded much of the Grand Coulee, and the part of the Spokane floods that came this way did the rest of the job.

When the glacier melted, the Columbia River returned to its old channel, and left the old stream channel and floodway without a flowing stream. U.S. 2 crosses Grand Coulee at the dividing point between upper and lower Grand Coulee, which extend north and south of the road, respectively. Dry Falls, about 3 miles southwest of Coulee City, is at the head of lower Grand Coulee.

Dry Falls

Dry Falls is unquestionably the most dramatic sight along the Grand Coulee, one of the most fascinating in the entire state. The thundering water with its spray full of rainbows is gone now, but everything else remains of what was in its time the world's greatest waterfall. Here, the Columbia River, at times augmented by the muddy Spokane floods, crashed over a precipice almost 400 feet high, and more than 3 miles wide. Niagara Falls would look puny if we could compare it to Dry Falls in its prime.

Like all large waterfalls, Dry Falls retreated upstream as the plunging river undercut its lip, and caused the lip of the falls to collapse. The retreat of Dry Falls left behind a deep gorge, which now remains as a dry valley, the lower Grand Coulee.

A small portion of the lip of Dry Falls. Imagine the torrents of the Spokane floods, yellow with silt eroded from the Palouse loess, thundering over this precipice with a roar that shook the ground for miles. Now, ground water fills the old plunge pools to make quiet little lakes.

Upper Grand Coulee

Upper Grand Coulee, north of U.S. 2, also formed as a gorge left in the path of a retreating waterfall. with some help from the bedrock structure. The old channel follows the hinge of a sharp bend in the basalt flows, which probably formed as they draped over a fault in the older rock beneath. Bending cracked the basalt into much smaller than normal pieces, making it much more easily erodible than the undeformed rock. The west wall of upper Grand Coulee stands much higher than the east side, because it is on the high side of the bend.

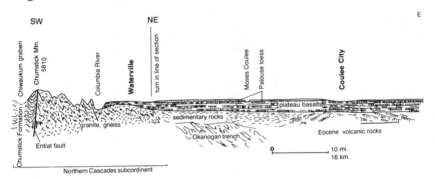

Section along the line of U.S. 2 in the area between the Columbia River and Coulee City showing the plateau basalt lapping onto older granitic rocks of the North Cascade subcontinent.

The waterfall that left upper Grand Coulee behind as it retreated vanished as its lip worked back into the older granite that lay below the basalt. Granite does not erode as easily as basalt, because it contains fewer fractures. Grand Coulee Dam, about 25 miles north of U.S. 2, rests on foundations in that granite, and impounds the Columbia River to form Franklin D. Roosevelt Lake. Large pumps transfer part of the water from that lake into upper Grand Coulee, to store it in Banks Lake, an equalizing reservoir that regulates flow over the dam. Dry Falls Dam at the south end of upper Grand Coulee prevents Banks Lake from going over the old waterfall — it keeps Dry Falls dry.

Glaciation

Glaciers covered most of the northern third of Washington when the last ice age was at its maximum. One major mass of ice filled the lowlands of the Okanogan Valley, and pushed south onto the northern margin of the Columbia Plateau. A lobe of that ice crossed the present course of the Columbia River, and reached almost as far south as the line of U.S. 2 between Orondo and Coulee City. That glacier left a large deposit of glacial till called the "Withrow moraine" to mark its

*The glacier plucked these giant boulders of basalt out of the north edge
of the Columbia Plateau, and dropped them on the Withrow moraine
beside Washington 174 west of Grand Coulee Dam. People who live
along the Withrow moraine call such boulders "haystack rocks."*

southern margin. A six mile side trip north on Washington 172 to
Withrow is an easy way to see the moraine. It is a broad ridge several
miles wide full of irregular hills and depressions, which extends from
Chelan southeast to the area just north of Coulee City.

Scablands

Between Almira and Davenport, U.S. 2 follows one of the major
channelways the Spokane floods eroded through the loess soil and
into the basalt beneath. In this area, productive fields are few, and
ragged outcrops of basalt common. The old flood channels are dry,
except for occasional ponds and lakes, and trees grow in them.
Nevertheless, it is easy to see the almost complete absence of soil in
this flood scrubbed landscape, and not too difficult to imagine the
rush of water that scoured across the landscape.

Mima Mounds, one of the mysteries of Washington geology. Millions of these low hummocks dot the prairies of Washington, most abundantly in areas near the limit of glaciation. The number of theories to explain their origin approximately equals the number of geologists who have studied them.

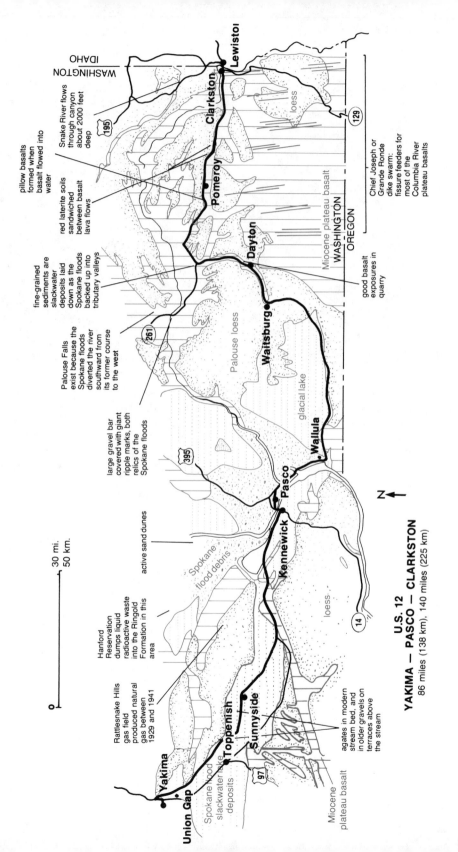

pillow basalts
formed when
basalt flowed into
water

Snake River flows
through canyon
about 2000 feet
deep

red laterite soils
sandwiched
between basalt
lava flows

Chief Joseph or
Grande Ronde
dike swarm:
fissure feeders for
most of the
Columbia River
plateau basalts

fine-grained
sediments are
slackwater
deposits laid
down as the
Spokane floods
backed up into
tributary valleys

good basalt
exposures in
quarry

Palouse Falls
exist because the
Spokane floods
diverted the river
southward from
its former course
to the west

Palouse loess

large gravel bar
covered with giant
ripple marks, both
relics of the
Spokane floods

glacial lake

active sand dunes

Spokane
flood debris

loess

Hanford
Reservation
dumps liquid
radioactive waste
into the Ringold
Formation in this
area

Rattlesnake Hills
gas field
produced natural
gas between
1929 and 1941

agates in modern
stream bed, and
in older gravels on
terraces above
the stream

Spokane flood
slackwater
deposits

Miocene
plateau basalt

Miocene plateau basalt
WASHINGTON
OREGON

WASHINGTON
IDAHO

Lewiston
Clarkston
Pomeroy
Dayton
Waitsburg
Wallula
Pasco
Kennewick
Toppenish
Sunnyside
Yakima
Union Gap

30 mi.
50 km.

N

U.S. 12
YAKIMA — PASCO — CLARKSTON
86 miles (138 km), 140 miles (225 km)

U.S. 12
Yakima—Pasco—Clarkston
86 mi. 138 km. 140 mi. 225 km.

All the bedrock exposed along U.S. 12 between Yakima and Clarkston is basalt lava flows of the Columbia Plateau, much of it buried beneath Palouse loess west of Wallula Gap, beneath Spokane flood deposits in the Yakima Valley and the Pasco basin. The rock is distinctive at a distance for its tendency to break into vertical columns, nearby for its very dark color, black washed with a rusty stain of iron oxide. These rocks erupted during late Miocene time, about 15 million years ago, give or take 2 or 3 million years.

South of Yakima, the highway passes through Union Gap, a narrow canyon that the Yakima River cut through a rising anticline in the Columbia Plateau. Ahtanum Ridge west of the highway and the Rattlesnake hills to the east are the same anticline. Between Union Gap and Granger, the road follows the north side of the Yakima Valley with Toppenish Ridge, another anticlinal arch in the basalt plateau, defining the southern skyline. Ahtanum Ridge and the Rattlesnake hills lie along the north side of the valley. Then the road follows the Yakima River east into the Pasco basin, another low area between anticlinal arches buckled into the lava plateau.

Dry Holes

Two unsuccessful deep wildcat wells were drilled close to the highway near the west end of Rattlesnake Ridge. Oil and gas are lighter than water, so they move up through the rocks until something blocks their way, commonly the crest of an anticlinal arch, because that is where the layers of rock that contain them reach their highest point. Petroleum is organic material that forms in sedimentary rocks that accumulated in sea water, not in basalt lava flows. Obviously, the wildcat wells were drilled in the belief that they would pass through the basalt, and penetrate an anticline in older sedimentary rocks beneath. However, neither well found sedimentary rocks. Further investigation showed that most of the basalt beneath Rattlesnake Ridge erupted during Eocene time, long before activity began in the Columbia Plateau.

If you draw a smooth curve on the map connecting the Eocene volcanic rocks in the Republic graben of northeastern Washington with volcanic rocks of the same age in the Ochoco Mountains of central Oregon, it passes close to Yakima. Perhaps the wildcat wells in the Rattlesnake Ridge found the connection between the Ochoco Mountains and the Republic graben. If so, a well drilled farther east would have a much better chance of finding sedimentary rocks beneath the Columbia Plateau.

The Wallula Gap

The Wallula Gap.

As the arch of the Horse Heaven Hills rose across the path of the Columbia River, the stream eroded its channel through the resistant basalt flows to cut the Wallula Gap. That was a slow process, and it backed the river up in the Pasco basin where it deposited several hundred feet of sediment called the Ringold formation, which consists largely of very fine grained material. The anticlinal ridges north and south of Pasco continue to grow, and the layers of the Ringold formation tilt up around the margins of the basin.

Flood Debris

These rippling beds of sand and silt are typical of those in the Touchet formation. They are sediment laid down in slackwater lakes that briefly existed as the Spokane floods backed up streams tributary to the Columbia River.

After the Ringold formation accumulated, layers of younger sediment covered it in much of the Pasco basin. Most of that material belongs to the Touchet formation, an extraordinary deposit that veneers the surfaces of low valleys all along the lower Columbia River. It gets its name from the Touchet River north of Walla Walla where some of the finest exposures of slackwater flood deposits exist.

The Touchet formation consists of debris dumped in slackwater lakes that existed for a few days at a time when the Spokane floods backed water into low areas along the Columbia River and its tributaries. In most exposures, the formation consists of numerous similar sequences of layers, as many as 40 in some places, which appear to record successive Spokane floods. Here and there, the fine sediments of the Touchet formation contain isolated boulders evidently dropped from icebergs that drifted in the temporary lakes. Some of the boulders are rocks that could only have come from western Montana, where the Spokane floods originated.

Highway 12 passes through this fascinating road cut about 8 miles east of its junction with U. S. 395. Bedrock exposed in the right side of the picture is gently dipping basalt lava flows. The left side of the picture shows an old channel filled with debris left by a Spokane flood. More flood debris in the top of the exposure blankets both the basalt and the old channel.

Radioactive Waste

Since 1945, the Atomic Energy Commission has been dumping radioactive waste water at the Hanford Reservation about 30 miles up the Columbia River from Pasco. Most of that water is still in the Ringold formation, where it moves very slowly because the material is so fine-grained. However, some has seeped into the overlying Touchet formation, which contains beds of coarse gravel, and therefore permits the radioactive material to move fairly rapidly.

Eventually, the radioactive waste water will soak through the Ringold formation, and into basalt lava flows below. Once in the basalt, the water may move considerable distances by percolating vertically along the fractures that define the columns in the lava flows, and horizontally through the rubbly zones between flows. No one seems to know where that pattern of flow will take the water, but there is some hope that the folds in the bedrock may confine it to the Pasco basin.

An enormous volume of liquid radioactive waste is now stored in the fine sands and silts of the Ringold formation beneath the Hanford Reservation. Under present conditions, most of that radioactive material will probably remain fairly safely where it is for a very long time. However, if the Columbia River were to shift its course and start rapidly eroding the soft silts which contain that radioactive matter, extremely serious contamination problems would extend to the mouth of the river. At this point, there is probably no alternative to simply hoping that no such thing happens. Nothing but time, thousands of years, can render that radioactive material harmless. There is no apparent way to get it out of the Ringold formation, nor any obviously better place to put it if we could get it out.

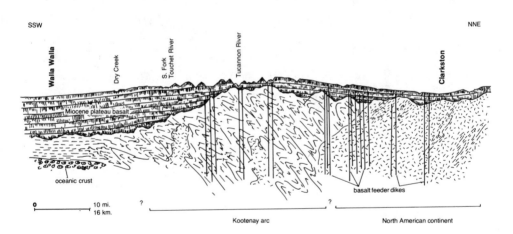

Section along a line between Clarkston and Walla Walla.

The Monocline

U.S. 12 follows the valley of the Snake River about 6 miles west of Clarkston through outstanding exposures of the plateau basalt. Very few areas offer better views of basalt columns, and none provides a better view of a spectacular warp in the Columbia Plateau. The layers of basalt tilt steeply down to the south in a simple fold of a type geologists call a monocline. This one drops the level of the plateau surface approximately 2000 feet from the north to the south side of the Snake River canyon. The basalt lava flows still lie nearly horizontally both north and south of the canyon, and slope from one elevation to the other through the monocline.

Long columns of basalt stand like rows of close-packed telephone poles in this quarry beside U.S. 12 about 6 miles west of Clarkston. A pale gray crust of calcite deposited from seeping ground water paints the column surfaces.

Palouse Loess

The Spokane floods passed north and west of the route of U.S. 12, between Wallula Gap and Clarkston, so the basalt along that part of the route still retains its blanket of Palouse loess. The rolling hills are old dunes of wind blown dust that marched across this landscape during times when the wind blew great clouds of dust. Because the hills have the form of dunes, it seems that those must also have been times when the climate was too dry to stabilize them beneath a protective cover of plants. Unfortunately, there is no obvious way to determine when that was.

U.S. 97
Columbia River — Ellensburg
112 mi. 181 km.

The route of U.S. 97 between the Columbia River and Yakima follows the western edge of the Columbia Plateau, where the flood basalts lap onto the Cascades, and younger volcanic rocks of the High Cascades lap onto the flood basalts. Along most of the route, the Cascades cut jagged skyline along the western horizon.

In this western margin of the Columbia Plateau, the northward movement of western North America is crumpling the lava flows into folds that trend generally from east to west, and make long ridges in the landscape. U.S. 97 crosses the trend of those folds in a series of great roller coaster ups and downs. Watch the layers of basalt exposed in roadcuts to see how they tilt north in some areas, south in others as they wrap over the folds.

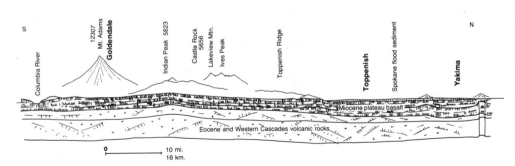

Section along the line of U.S. 97 between the Columbia River and Yakima.

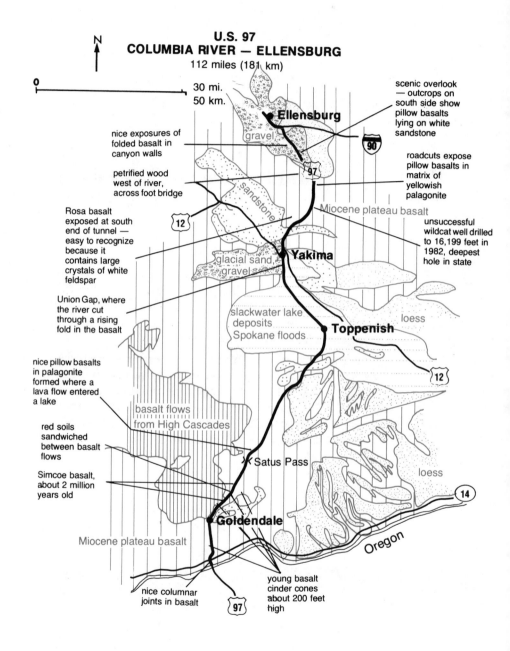

N

**U.S. 97
COLUMBIA RIVER — ELLENSBURG**
112 miles (181 km)

0

30 mi.
50 km.

scenic overlook
— outcrops on
south side show
pillow basalts
lying on white
sandstone

● **Ellensburg**

gravel

90

nice exposures of
folded basalt in
canyon walls

97

roadcuts expose
pillow basalts in
matrix of
yellowish
palagonite

petrified wood
west of river,
across foot bridge

sandstone

Miocene plateau basalt

Rosa basalt
exposed at south
end of tunnel —
easy to recognize
because it
contains large
crystals of white
feldspar

12

unsuccessful
wildcat well drilled
to 16,199 feet in
1982, deepest
hole in state

glacial sand,
gravel

Yakima

Union Gap, where
the river cut
through a rising
fold in the basalt

slackwater lake
deposits
Spokane floods

loess

Toppenish

nice pillow basalts
in palagonite
formed where a
lava flow entered
a lake

12

red soils
sandwiched
between basalt
flows

basalt flows
from High Cascades

Simcoe basalt,
about 2 million
years old

Status Pass

loess

14

Goldendale

Miocene plateau basalt

Oregon

nice columnar
joints in basalt

97

young basalt
cinder cones
about 200 feet

The Flood Basalts

All of the flood basalts in this area erupted from the Grande Ronde volcano near the corner of Washington, Oregon, and Idaho during late Miocene time. Geologists who specialize in the Columbia Plateau have distinguished and named most of the individual flows by learning to recognize small differences in the look of the rock, and by comparing their chemical compositions. How they manage to recognize individual lava flows is no less a mystery to most other geologists than to people who are not geologists. One flood basalt flow really does look very much like another.

Nevertheless, the flood basalt flows do contain a number of interesting features. One of the commonest complexities formed in places where a lava flow poured into a lake. When that happens, the lava flow congeals into pillow basalt, and the lake boils. The hot basalt and steam react with each other to form a yellowish rock full of clay, palagonite. The end result is a wild roadcut full of broken pillows of basalt set in a matrix of yellowish palagonite, which is usually mixed with sand and mud from the bottom of the lake. Pillow and palagonite complexes appear in every part of the Columbia Plateau; watch for an especially nice one in a roadcut 3 miles north of Satus Pass.

Young Basalt

About 3 miles south of Goldendale, U.S. 97 crosses a basalt lava flow that erupted from Lorena Butte, a small cinder cone volcano visible about 3 miles east of the road. The eruption happened early in Pleistocene time, sometime more than a million years ago.

North of Goldendale, the highway follows the valley of the Little Klickitat River to its headwaters just south of Satus Pass. Along the southern half of that 14 mile route, the highway passes much more young basalt, and the evidence of its age is clearly exposed in several places. Watch for exposures of dark basalt lying on top of beds of gravel. The gravel belongs to the Ellensburg formation, which accumulated on top of the flood basalt during latest Miocene and Pliocene time, when the climate in this region was very dry. The Ellensburg formation was still accumulating as recently as 2 million years ago, so we can be sure that the basalt on top of it is younger than that.

Faults and Folds

About one and one-half miles north of the junction with Washington 14, U.S. 97 crosses the Wishram fault, which trends from east to west, and dropped the area to the south. Look for several roadcuts on the west side of the highway in broken basalt crushed by movement on the fault, a fault breccia.

Satus Pass is at the crest of the Horse Heaven Hills, an anticlinal arch in the Columbia Plateau. Toppenish Ridge is another such fold, which the road also crosses. North of Toppenish Ridge, U.S. 97 crosses a vast plain that extends in all directions from Toppenish. This is the Toppenish basin, a synclinal trough between anticlinal arches. Its floor is so flat because it contains a deep fill of young sedimentary deposits, including debris dumped from temporary lakes that filled the basin while the Spokane floods backed behind the bottleneck of Wallula Gap.

Rattlesnake Ridge along the north side of the Toppenish basin is an anticlinal arch, which the Yakima River cut through as it rose to form Union Gap, an easy low path for the highway. Just north of Yakima, the road passes through another narrow gorge, Selah Gap, which the Yakima River cut through another rising anticlinal arch. Between Selah Gap and Ellensburg, the river passes through a narrow canyon it eroded through the Umtanum and Manastash ridges, two more anticlinal arches that rose across the path of the stream. The old highway follows that canyon past spectacular cliff exposures of folded basalt. However, the modern highway crosses the top, passing through many roadcuts that provide almost equally good exposures.

Steeply tilted lava flows and white sedimentary rock sandwiched between them exposed in a roadcut on Manastash Ridge, about 9 miles south of Ellensburg.

U.S. 195
Spokane — Lewiston-Clarkston
100 mi. 160 km.

All the bedrock along the road between Spokane and Clarkston is basalt lava flows of the Columbia Plateau. Good exposures abound in the scablands and the larger stream valleys, but deep deposits of Palouse loess almost completely blanket the plateau surface. Several prominent buttes east of the highway are old hills of the Northern Rocky Mountains that stand like islands above the sea of basalt lava flows.

Fossil Tundra

Between Spokane and Spangle, the highway crosses the Cheney-Palouse tract of channeled scablands, one of the major water courses of the Spokane floods. Most of the Palouse loess is gone in this area, except on the tops of the higher hills, and there are large exposures of flood scrubbed basalt. The area also contains deposits of glacial outwash shed from melting glaciers at the end of the last ice age, and Glacial Lake Spokane left its own sedimentary record. About a mile north of Spangle, right beside the highway, there is an outstanding area of polygonal ground developed mostly on glacial outwash.

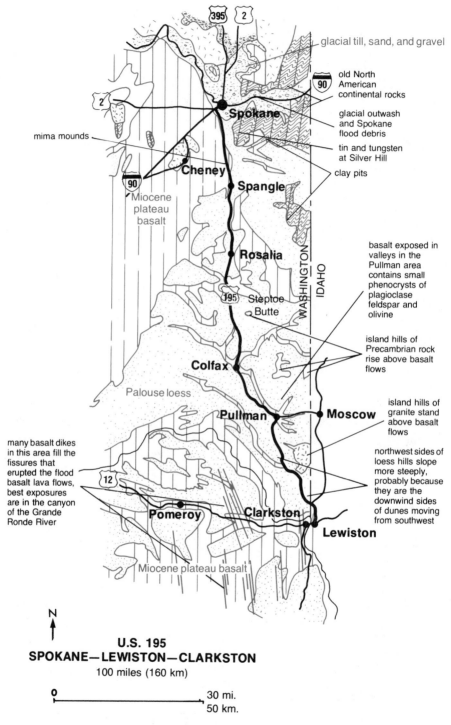

glacial till, sand, and gravel

old North
American
continental rocks

glacial outwash
and Spokane
flood debris

tin and tungsten
at Silver Hill

clay pits

mima mounds

Spokane

Cheney

Spangle

Miocene
plateau
basalt

Rosalia

Steptoe
Butte

basalt exposed in
valleys in the
Pullman area
contains small
phenocrysts of
plagioclase
feldspar and
olivine

island hills of
Precambrian rock
rise above basalt
flows

Colfax

Palouse loess

Pullman

Moscow

island hills of
granite stand
above basalt
flows

many basalt dikes
in this area fill the
fissures that
erupted the flood
basalt lava flows,
best exposures
are in the canyon
of the Grande
Ronde River

northwest sides of
loess hills slope
more steeply,
probably because
they are the
downwind sides
of dunes moving
from southwest

Pomeroy

Clarkston

Lewiston

Miocene plateau basalt

WASHINGTON

IDAHO

N

U.S. 195
SPOKANE—LEWISTON—CLARKSTON
100 miles (160 km)

0   30 mi.
                     50 km.

Polygonal ground forms today in tundra regions of the high arctic and many high mountains where the subsoil is permanently frozen. Freezing and thawing of the top level of the soil produces a pattern of low mounds a foot or so high and 10 to 30 feet across evenly spaced on the ground surface. Narrow zones of coarser rock outline each mound, and connect to make a continuous net covering the entire surface.

Old polygonal ground exists in many parts of the Columbia Plateau where the Palouse loess does not deeply blanket the bedrock. It typically appears as narrow stripes of black chunks of basalt, which contrast sharply with the yellow grass. They are fossil tundras, which tell us that the ice age climate was considerably colder than the one we know, probably cold enough to keep the subsoil permanently frozen.

Loess

Many roadcuts in the Palouse Hills south of Spangle expose wind blown dust, the Palouse loess. It lies in deposits as much as 200 feet deep, and in hills about that high, that cover the basalt surface. Roadcuts expose interesting patterns of thin layers in the loess, which probably record movement of the dust dunes that became the Palouse Hills. Although this must have been a bleak setting when the wind was blowing those dust dunes across the plateau surface, the loess now makes marvelously deep and fertile agricultural soils.

Steptoe Butte

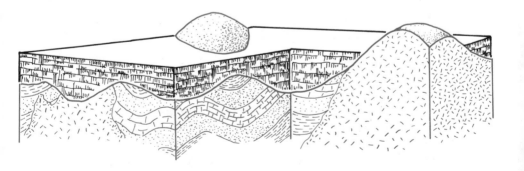

Old hills rising above the plateau surface show that the Columbia Plateau basalts bury old continental rock along the eastern fringe of Washington.

The highway follows a route close to the margin of the Columbia Plateau, where the flood basalt flows backed up into the valleys of the Northern Rocky Mountains, but did not cover the hills. Higher hills that rise above the plateau surface between Spokane and Clarkston are outposts of the Northern Rockies. The western limit of their distribution approximately marks the western edge of the old North American continent.

Steptoe Butte, just northeast of the little community of Steptoe, is the best known of those ancient island hills that stood here before the Columbia Plateau formed. It consists of billion year old Precambrian Belt sedimentary rocks. Steptoe Butte is a state park, and a side road east of U.S. 195 from the town of Steptoe, spirals to its top, where the view is wonderful. The Columbia Plateau spreads away to the west, like a floor under the sky, and the older mountains of Idaho march across the eastern horizon. Kamiak Butte, a larger version of Steptoe Butte, rises above the plateau surface as a conspicuous ridge about 10 miles north of Pullman. Any prominent butte along this road has a similar story.

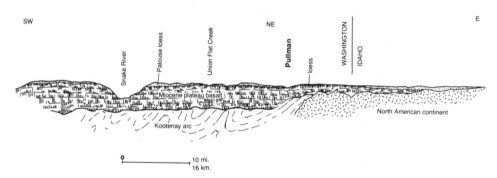

Section across the line of the highway near Pullman. The flood basalts lap onto the Northern Rocky Mountains.

Basalt

The best exposures of basalt are near Pullman and Colfax, where the road dips into the valley of the Palouse River, and around Clarkston, where the Snake River flows through a spectacular gorge eroded into the Columbia Plateau. As everywhere, basalt is unmistakable, a black rock that contains very few or no crystals large enough to be visible without a microscope. Rusty iron oxides stain most of the basalt on the Columbia Plateau brown, so the black color appears mostly on freshly-broken surfaces.

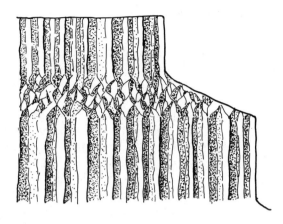

Thick lava flows have colonnades of columns at their top and base, a zone of irregularly fractured basalt between.

Basalt tends to break into vertical columns along fractures that form as the lava flow shrinks slightly as it crystallizes. Ledges exposed in canyon walls, such as those of the deep canyon of the Snake River around Clarkston, suggest stockades of black logs standing vertically in the cliff face. Only relatively thin lava flows break into a single row of columns. Thicker basalt flows, and most of those in the Columbia Plateau are thick, tend to break into two colonnades of columns, one at the top and another at the base, with a zone of irregularly fractured basalt separating them.

The Bend in the Basalt

U.S. 195 crosses into Idaho just a few miles north of Lewiston. The scenic overlook about 2 miles south of the state line is near the hinge of a bend that warps the Columbia Plateau down to the south as though it were a magazine draped over the edge of a book. The basalt that forms the plateau surface around the scenic overlook is the same as that exposed in the plateau surface visible in the distance south of the river, where it is about 2000 feet lower.

Between Lewiston and the scenic overlook, the road crosses a canyon wall about 1800 feet high in a series of sweeping switchback curves. Roadcuts provide numerous good exposures of basalt, and those who can leave the driving to others may notice that many of the layers are tilted down to the south. That tilt reflects the fold that warps the plateau surface down about 2000 feet here.

Snake River Floods

The Snake River has had more than its share of adventures with catastrophic floods, all during the last ice age. They began with the Bonneville flood. It happened when Lake Bonneville, a large body of water that deeply flooded the basin of the Great Salt Lake and covered a large area of Utah overflowed through Red Rock Pass south of Pocatello, Idaho. The rock in the top of the pass eroded very easily, so the lake level dropped rapidly, and water poured down the valley of Marsh Creek, past Pocatello, and into the Snake River in enormous volume. That went on until the water had cut the pass down some 375 feet into hard rock, which stabilized the level of the lake. By that time, floodwaters had scoured the entire valley of the Snake River, creating features much like those left by the Spokane flood, but on a much smaller scale.

The Spokane floods came later, and they did not pour down the Snake River. Instead, they reversed its direction of flow, and backed it up to a maximum depth estimated to have been about 600 feet at Lewiston. At least 20 distinct layers of backwater deposits in the Lewiston area record the Spokane floods that backed water this far up the Snake River. Evidently, some of the lesser Spokane floods did not affect the Snake River this far upstream.

U.S. 395
Pasco — Ritzville
74mi. 118 km.

Between Pasco and its junction with I-90 at Ritzville, U.S. 395 crosses basalt lava flows of the Columbia Plateau, including some of the youngest flows. However, deep soils, sand dunes, and fertile deposits of wind blown dust, the Palouse loess, cover the basalt along most of the route. Exposures of basalt appear only in the scabland channels.

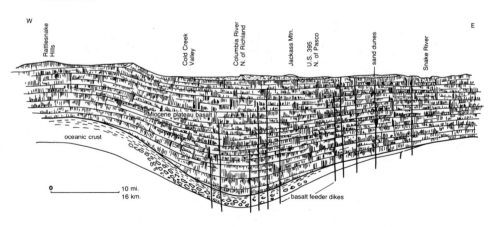

Section across the line of U.S. 395 between Pasco and Eltopia.

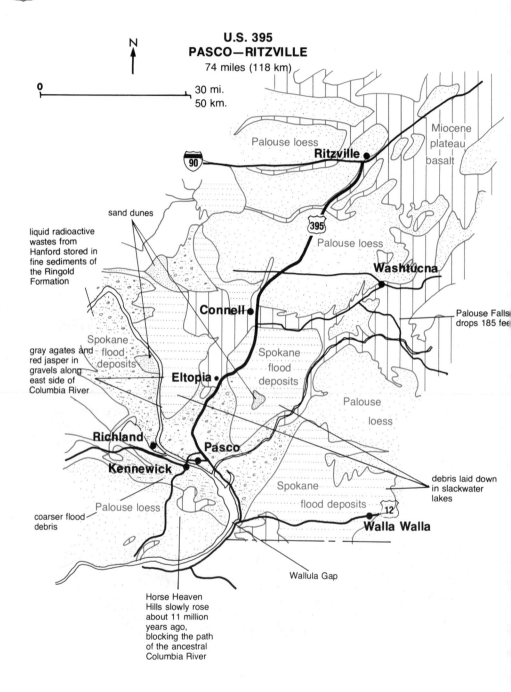

U.S. 395
PASCO—RITZVILLE
74 miles (118 km)

N

0   30 mi.
         50 km.

Palouse loess

Ritzville

Miocene
plateau
basalt

90

395

sand dunes

Palouse loess

liquid radioactive
wastes from
Hanford stored in
fine sediments of
the Ringold
Formation

Washtucna

Connell

Palouse Falls
drops 185 feet

Spokane
flood
deposits

gray agates and
red jasper in
gravels along
east side of
Columbia River

Spokane
flood
deposits

Eltopia

Palouse
loess

Richland

Pasco

Kennewick

debris laid down
in slackwater
lakes

Spokane
flood deposits

12

Walla Walla

Palouse loess

coarser flood
debris

Wallula Gap

Horse Heaven
Hills slowly rose
about 11 million
years ago,
blocking the path
of the ancestral
Columbia River

The Pasco Basin

Pasco, Richland, and Kennewick are in the Pasco basin, a broad trough warped into the Columbia Plateau. The floor of the Pasco basin is very flat, because it contains a deep fill of sediments, many of them left there when the Spokane flood ponded behind the bottleneck of the Wallula Gap. Now the tireless wind that whips across this open expanse of dry country has blown some of that sediment into dunes.

Sand Dunes

Wind ripples and little footprints in the soft surface of a sand dune.

The area around and east of Eltopia contains the finest field of sand dunes in Washington, the Juniper Dunes. They cover dozens of square miles. Some of the dunes now support enough plant cover to protect them from the wind, but most continue to move despite scattered patches of vegetation. In large parts of the area, the only conspicuous plants are gnarled little juniper trees, natural bonsais that look extremely picturesque among the dunes.

Although most people imagine sand dunes when they think of deserts, most deserts contain relatively few dunes, and those typically gather into patches, like the Juniper Dunes. That happens because dunes have soft surfaces. Blowing sand grains tend not to bounce off the soft surface of a dune for the same reason that a golf ball tends not to bounce out of a patch of tall grass. So dunes grow as they collect passing sand grains, and the wind sweeps the sand of the desert up into neat piles, instead of scattering it over the countryside.

If the wind whips sand across the flat surface of the Pasco basin, it must also move dust. But unlike sand, which skips in little bounces across the ground surface, dust blows high, and moves long distances. So it is hard to know where the dust that blows out of the Pasco basin goes, but at least some of it must be in the Palouse loess of eastern Washington.

Between Scablands

Long stretches of U.S 395 between Pasco and Ritzville follow high ground, and would have survived unscathed had they existed when the Spokane floods came through. The monster floods scrubbed the soil and loess off large areas east and west of the road, leaving ragged tracts of exposed basalt. However, the highway does cross a tract of flood-scoured scabland just south of Connell, and between Lind and Ritzville. Bedrock along the road in those areas is basalt.

The wind blown dust of the Palouse loess makes extremely fertile soils that nourish abundant crops, such as this new growth of winter wheat.

Washington 14
Vancouver—Pasco
201 mi. 320 km.

Along most of the route, Washington 14 follows the bluffs north of the Columbia River passing from one spectacular scene to the next. Long stretches of the road provide magnificent views of Oregon's Mount Hood cutting a jagged piece out of the sky south of the river. Mount Hood, a great snowy monster of a volcano, has lost most of its original volcanic form and surface to glaciation, and so has the look of a volcano that has not erupted for a very long time. Nevertheless, there are several fairly believable accounts of activity during the middle of the last century, and geologic evidence of eruptions within the last few thousand years. No geologist would be surprised to see it erupt.

All the roadcuts and outcrops along the route expose volcanic rock. Basalt flows erupted from the Cascade volcanoes appear in the Columbia Gorge between Vancouver and the Dalles. East of the Dalles, all of the bedrock is Columbia Plateau basalt.

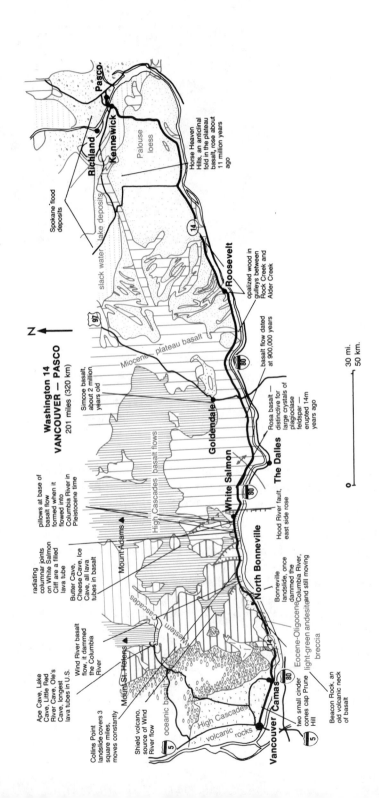

Washington 14
VANCOUVER — PASCO
201 miles (320 km)

N

Spokane 'flood deposits

Palouse loess

Horse Heaven Hills, an anticlinal fold in the plateau basalt, rose about 11 million years ago

slack water lake deposits

opalized wood in gulleys between Rock Creek and Alder Creek

Roosevelt

Miocene plateau basalt

basalt flow dated at 900,000 years

Simcoe basalt, about 2 million years old

Rosa basalt — distinctive for large crystals of plagioclase feldspar — erupted 14m years ago

Goldendale

High Cascades basalt flows

White Salmon

The Dalles

Hood River fault, east side rose

30 mi.

50 km.

pillows at base of basalt flow formed when it flowed into Columbia River in Pleistocene time

radiating columnar joints on White Salmon Cliff are a filled lava tube

Butter Cave, Cheese Cave, ice tubes in basalt

Mount Adams

Bonneville landslide, once dammed the Columbia River, and still moving

North Bonneville

Eocene-Oligocene light-green andesite breccia

Ape Cave, Lake Cave, Little Red River Cave, Ole's Cave, longest lava tubes in U.S.

Wind River basalt flow that dammed the Columbia River

Collins Point landslide covers 3 square miles, moves constantly

Shield volcano, source of Wind River flow

Mount St. Helens

oceanic basalt

Western Cascades

High Cascades volcanic rocks

Vancouver/Camas

two small cinder cones cap Prune Hill

Beacon Rock, an old volcanic neck of basalt

Richland

Pasco

Kennewick

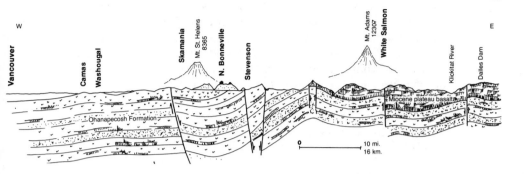

Section along the Columbia River between Vancouver and the Dalles.

The Cascade Arch and the Columbia Gorge

The Cascade volcanoes stand on the crest of an anticline in the plateau basalt. That fold arches the Columbia Plateau basalts, so it must have formed since they erupted. It probably reflects thermal expansion of the rocks along the trend of the Cascade volcanoes.

The Columbia River eroded through the rocks as fast as the Cascade arch rose, thus maintaining its course, and cutting the Columbia Gorge. Early settlers found the Columbia Gorge with its numerous cascades and extremely narrow valley floor so nearly impassable that many followed an alternate route south of Mount Hood. Now the river lies quietly impounded behind a dam, and streams of traffic pour through the narrow valley on highway, railroads, and river boats.

Western Cascade Volcanic Rocks

In the area between North Bonneville and White Salmon, the river eroded through the Columbia Plateau lava flows into older volcanic rocks beneath, the Ohanepecosh formation of the Western Cascades. They consist of basalt and dark andesite erupted during late Eocene and Oligocene time, between 40 and 20 million years ago. The formation was named for exposures near Mount Rainier.

Lava Dams

Lava flows erupted from the modern Cascade volcanoes dammed the Columbia Gorge at least twice. On one occasion, lava poured south down the Wind River to impound the Columbia River about one

215

mile east of Stevenson. No one knows how long the lake lasted before the Columbia River removed the lava dam, but we can see the remains of a delta at least 150 feet thick near Stevenson. Attempts to obtain a radiocarbon date on a log found buried in the delta gave an age greater than 35,000 years, the maximum that the radiocarbon method can measure. So we know only that the lake filled sometime before 35,000 years ago, but not how long before. A much larger lava dam formed in the Hood River area when flows poured into the Columbia Gorge from both sides. The age of this lake is likewise something more than 35,000 years.

Giant Landslide

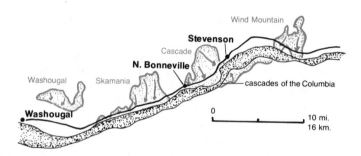

Some of the larger landslides in the Columbia Gorge.

Several enormous landslides exist on the Washington side of the river where it cut through the Columbia Plateau basalts into the older volcanic rocks of the Ohanepecosh formation. In this area, the layers of rock slope gently downward to the south, so they would naturally tend to slide towards the river. The sequence and arrangement of the rocks greatly encourage that tendency.

The uppermost rocks are the plateau basalt flows, which are full of vertical shrinkage fractures formed as the lava crystallized. Those are the fractures that so commonly make exposures of basalt break into colonnades of vertical columns. Rainwater soaks down through those fractures into a thick layer of ancient soil buried beneath the lava flows. The soil holds the water because the older volcanic rocks beneath are impermeable. You could simulate the effect of the wet soil sandwiched between the Ohanepecosh formation below and the Columbia Plateau basalts above by spreading a layer of grease on a sloping board, and then laying another board on top. It is no surprise to find the upper basalt flows sliding downslope into the Columbia River on the water soaked layer of buried soil.

Washington 14 crosses the Cascade slide in the 5 miles between North Bonneville and Stevenson. It is almost too big to see from the ground, but perfectly obvious from the air. The Cascade slide once crossed the Columbia River to form a dam that soon washed out leaving a few boulders to form the Cascades of the Columbia, now submerged beneath the waters behind Bonneville Dam. The Cascade slide still crowds the Columbia River against the Oregon shore, to form the narrows in which Bonneville Dam was built.

Lewis and Clark noted numerous dead trees that had drowned in the temporary lake behind the Cascade slide still standing as tall snags in the forest along the shore. Radiocarbon dates on wood from those trees show that they died during the thirteenth century. So it is reasonable to suppose that the Cascade slide inspired the Indian legend of "The Bridge of the Gods."

Of the several smaller slides, the Wind Mountain slide, about 6 miles east of Stevenson, is probably the most active. It now moves between 40 and 50 feet each year. That keeps the highway in constant need of repair, and also crowds the river against its south bank.

Spokane Floods

The Spokane floods came this way, mostly through the Wallula Gap, where the water ponded in the Pasco basin poured into the Columbia River. Geologists estimate that something in the neighborhood of 40 cubic miles of water per day poured through the Wallula Gap into the Columbia Gap during the Spokane floods. That is more than the current combined discharge of all the rivers of the world, approximately 80 times the maximum recorded flood discharge of the Mississippi River.

Passage of the Spokane floods left no soil cover on the basalt cliffs in the Wallula Gap.

That rush of water ponded again in broad reaches of the valley upstream from every narrows to form a series of temporary lakes. Deposits of sediment in those places record each ponded Spokane flood. Water rushing through the narrows scoured the soil off the valley walls, and the sediment from the valley floors. Watch below the road for terraces just above river level thickly strewn with large boulders, or jagged with outcrops of basalt. The best of those scoured surfaces was at the Dalles and Celilo Falls where the river flowed through deep channels in the basalt bedrock. Both the channels and the falls are now flooded behind the Dalles Dam. Watch also above the road for scrubbed valley walls only scantily clad in soil to an elevation of about 1000 feet everywhere upstream from the Columbia Gorge. Above 1000 feet, outcrops of basalt become scarce, and the hills retain the smoothly upholstered look of slopes well-covered with soil.

Here and there, boulders of distinctive rocks imported from northern Idaho and western Montana litter the scrubbed valley walls below an elevation of 1000 feet. Watch for large chunks of gray granite and pink or red mudstone, none of which resemble the basalt bedrock. Most of those boulders are much too angular to have rolled all the way from Montana, so something must have carried them. They could only have arrived where we now find them embedded in floating icebergs that went aground on the hillslope. Similar ice rafted boulders also litter the Willamette Valley in Oregon, to show that the Spokane floods backed up there too.

View north across the Columbia River at Biggs Junction. The Spokane floods scrubbed soil off the valley walls and floor to reveal the basalt. The higher hills that stood above the floods still retain their cover of soil.

Washington 17
Coulee City — Moses Lake
50 mi. 80 km.

Between Coulee City and Soap Lake, the road follows the lower
Grand Coulee past marvelous cliff exposures of Columbia Plateau
basalts, and four lakes that flood low places in the floor of the old
channel. The route from Soap Lake through Ephrata to Moses Lake
crosses the Quincy basin, in which the Spokane floods left a deep fill of
debris.

*A close view of basalt exposed beside the road just south of Dry Falls.
The bubbles are gas cavities, which are generally common in basalt,
but rare on the Columbia Plateau. Probably this flow picked up a
charge of steam as it poured across marshy ground downstream from
Dry Falls.*

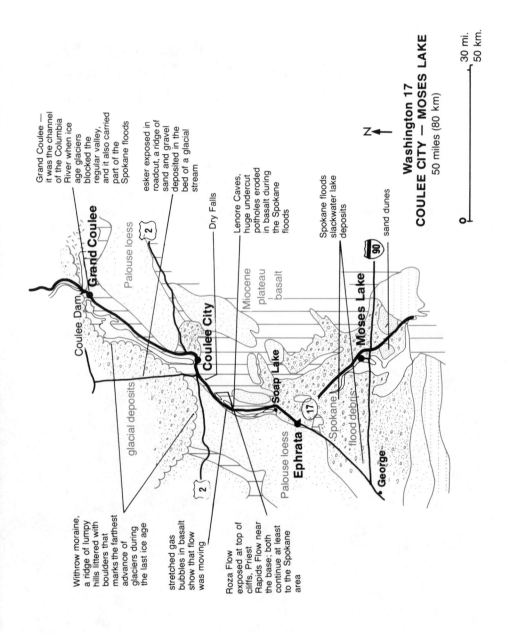

Grand Coulee — it was the channel of the Columbia River when ice age glaciers blocked the regular valley, and it also carried part of the Spokane floods

esker exposed in roadcut, a ridge of sand and gravel deposited in the bed of a glacial stream

Dry Falls

Lenore Caves, huge undercut potholes eroded in basalt during the Spokane floods

Spokane floods slackwater lake deposits

sand dunes

Grand Coulee

Palouse loess

Coulee Dam

Coulee City

Miocene plateau basalt

Soap Lake

Moses Lake

glacial deposits

Spokane flood debris

Ephrata

Palouse loess

George

Withrow moraine, a ridge of lumpy hills littered with boulders that marks the farthest advance of glaciers during the last ice age

stretched gas bubbles in basalt show that flow was moving

Roza Flow exposed at top of cliffs, Priest Rapids Flow near the base; both continue at least to the Spokane area

Washington 17
COULEE CITY — MOSES LAKE
50 miles (80 km)

N

0   30 mi.
         50 km.

The Dead Rhinocerus in Sun Lake State Park

Sun Lake State Park is east of the highway, a short distance south of Coulee City. Besides being in the midst of the geologic wonders of the Dry Falls area, the park contains one of the most extraordinary fossils ever found — the mold in basalt of a dead rhinocerus. Evidently, this rhinocerus was lying dead and bloated in a pond, when one of the Columbia Plateau lava floods filled the pond, and entombed the rhinocerus in basalt. Erosion of lower Grand Coulee cut into one leg just enough to expose the fossil as a small cave in the base of a lava flow. When some boys discovered the fossil in 1935, the cave still contained bones and teeth of the rhinocerus.

The Chain of Lakes

The highway passes a series of small and large lakes as it passes through the lower Grand Coulee. All of the lake basins probably formed as plunge pools below the lip of the retreating falls, so they become younger northward.

Basalt lava flows exposed in the cliff walls of Grand Coulee between Dry Falls and Soap Lake. Basalt columns always stand perpendicular to the cooling surface. The vertical columns in the right side of the picture formed in basalt that covered level ground, the wild columns farther left probably formed in basalt that cooled inward from the walls of an old channel.

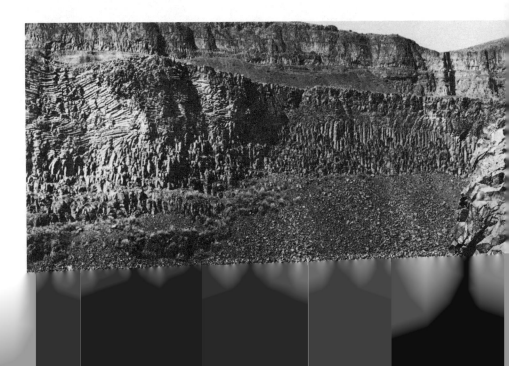

There is no better place than the drive along the lower Grand Coulee to see how basalt lava flows tend to break into vertical columns, which make the cliff exposures look like stockades. The Spokane floods were able to erode their deep channels in a short time because it is easy to pick those flows apart, column by column. Had the Spokane floods poured across almost any other kind of hard bedrock, the channeled scablands would be a lot less channeled.

Soap Lake

The chain of lakes in the lower Grand Coulee drains south, one into the other, into Soap Lake, which has no outlet except through evaporation. This is a dry region.

Soap Lake is one of several extremely unpleasant bodies of water in the dry part of Washington. It is slightly saltier than the ocean, and it also contains enough dissolved sodium carbonate to make the water distinctly alkaline. That is why it feels soapy, and why it tastes bitter, as well as salty. On windy days, big masses of foam that look like dirty soap suds pile up on the shores of the lake. The foam is surprisingly stiff and durable, supposedly because it contains a bit of oil. Many people believe that water from Soap Lake has medicinal value. We don't agree. Water from undrained lakes is rarely wholesome, and that from Soap Lake would be approximately equivalent to a salty solution of sodium bicarbonate.

The Quincy Basin

Most of the route between Soap Lake and Moses Lake crosses the Quincy basin, a synclinal trough in the folded Columbia Plateau. The Spokane floodwaters slowed down to make a sort of fast draining lake as they entered this broad basin, and therefore dumped large quantities of sediment. The deep fill of sediment completely buries the basalt bedrock, and also holds large quantities of ground water, now widely used for irrigation.

Most of the floodwater drained through the Drumheller channels south of the Potholes Reservoir into the Othello basin, where it ponded again to make another temporary lake. That water finally emptied into the Pasco basin by way of the Othello channels a few miles southwest of the Scootenay Reservoir. The rest of the floodwater drained out of the Quincy basin into the Columbia River through the Frenchman Springs and Potholes coulees near I-90 between George and Vantage.

Sand Dunes

This is dry country, desert for all practical purposes, and several tracts of sand dunes migrate across the Quincy basin, especially the southern part. Most of the dunes are the kind that geologists call "barchans," which have a distinctive crescent shape. Dunes of that type form in places where winds strong enough to move sand blow consistently from the same direction. In the Quincy basin, the barchan dunes migrate generally eastward, with the horns of the crescent pointing the way.

One tract of sand dunes forms a natural dam that impounds Moses Lake. Potholes Reservoir, the body of water impounded by the O'Sullivan Dam south of the town of Moses Lake, also raised the level of Moses Lake, and flooded a tract of sand dunes. The crests of some of those dunes now stand as many small islands in the lake. The wind is gradually blowing the dry tops off those islands, and will eventually grade all of them down to the level where the sand becomes too moist and sticky to move.

Scablands

South of Moses Lake, between Othello and Eltopia, Washington 17 crosses scablands. Floodwaters entered this area from the north through the Othello channels, and from the east through a channel that follows the line of Washington 260 east to Connell. The water passed through on its way to the Pasco basin, where it ponded to form a temporary lake.

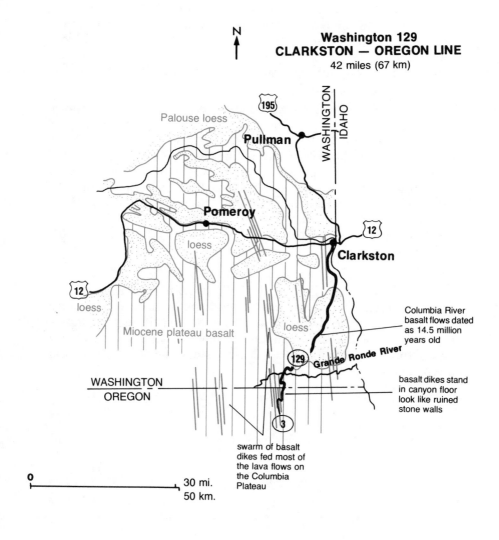

N

Washington 129
CLARKSTON — OREGON LINE
42 miles (67 km)

195

Palouse loess

WASHINGTON | IDAHO

Pullman

Pomeroy

12

loess

Clarkston

12

loess

loess

Miocene plateau basalt

Columbia River
basalt flows dated
as 14.5 million
years old

129 Grande Ronde River

WASHINGTON
OREGON

basalt dikes stand
in canyon floor
look like ruined
stone walls

3

swarm of basalt
dikes fed most of
the lava flows on
the Columbia
Plateau

0
30 mi.
50 km.

Washington 129
Clarkston — Oregon Line
42 mi. 67 km.

The road south from Clarkston follows a gently rising slope out of the Snake River Canyon, and onto the original surface of the Columbia volcanic plateau, mostly covered in this area by a rather open pine forest. Just north of the Oregon state line, the road descends into the deep canyon of the Grande Ronde River by way of a long succession of alarming switchbacks, a genuine adventure in motoring. Most people find a pause to look at the rocks entirely welcome.

This area on the flank of the big Grande Ronde volcano is still one of the higher parts of the Columbia Plateau. The flows are relatively thin in this area because they were pouring down the slope of the volcano on their way to flatter areas, where they ponded to make the thicker flows we more commonly see. Conspicuous red laterite soils sandwiched between the lava flows make it easy to judge their thickness, and also show that time enough for soil to develop passed between eruptions.

The Grande Ronde Dike Swarm

The bottom of the canyon provides one of the best views of the basalt dike swarm under the Grande Ronde volcanoes. Several large basalt dikes stand up in erosional relief like great ruined walls rising above the canyon floor. Numerous others appear on the canyon walls as black streaks cutting vertical paths through the nearly horizontal lava flows. Each dike is an old fissure now filled with solid basalt, the plumbing of a great lava flow.

South of the Grande Ronde Canyon, Oregon 3 climbs another long series of hair raising switchbacks back onto the plateau surface. Several unpaved side roads head east to a series of small buttes, each a small cinder cone volcano marking where one of the dikes broke the surface to become a lava flow. Harl Butte is one of the better and more easily accessible examples. Radioactive age dates on rocks from those buttes show that they are approximately 15 million years old. Those buttes mark the fissures that produced some of the last lava flows on the Columbia Plateau.

Lava flows make prominent ledges above Washington 129 in the canyon of the Grande Ronde River.

VI
THE WILLAPA HILLS

The Willapa Hills

The Willapa Hills roll gently through the southwestern corner of Washington. Bedrock throughout the region is simply a large slab of oceanic crust still lying almost as flat as it formed, but about two miles above the normal elevation of oceanic crust. The oldest rocks are greenish black pillow basalts, which erupted far offshore, and once formed the bedrock crust beneath the Pacific Ocean. In large areas, a deep cover of oceanic sedimentary rocks still blankets the pillow basalts.

Docking the North Cascade micro-continent established the oceanic trench in essentially its present position along a line offshore from Vancouver Island south to the coast of California. That left a large tract of oceanic crust, the Willapa Hills, stranded between the new trench and the west coast of the North Cascade subcontinent. That expanse of oceanic crust extends south through the Oregon Coast Range to the Klamath Mountains.

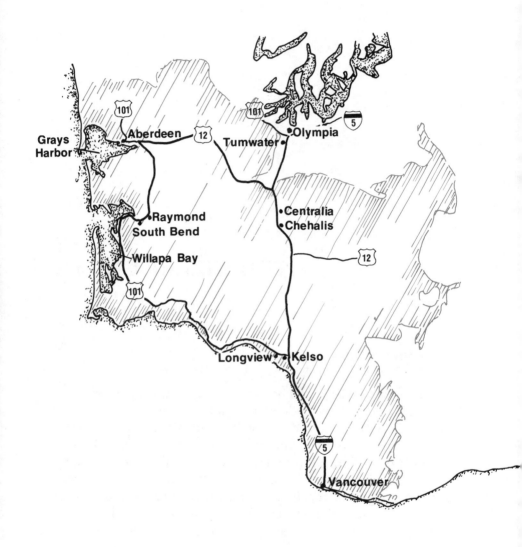

The Willapa Hills.

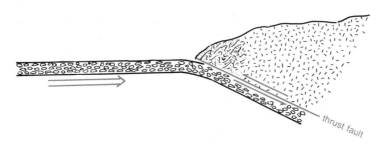

The oceanic crust of the Willapa Hills was shoved under the western edge of the North Cascade subcontinent.

In many areas, the old oceanic crust of western Washington meets the North Cascade subcontinent along faults, most notably the Shuksan thrust fault. They formed as the oceanic crust was shoved under the western edge of the North Cascade subcontinent after it docked. Along most of the eastern margin of the Willapa Hills, younger volcanic rocks of the Western Cascades cover the contact between the oceanic crust and the North Cascade subcontinent, obscuring whatever faults may exist there.

Oceanic sedimentary rocks that cover the pillow basalts in western Washington contain abundant fossils, a few large enough to see easily, most microscopic. The oldest rocks, those that rest directly on the pillow basalt, contain remains of animals that lived during Eocene time, probably sometime between 55 and 40 million years ago. That must have been about when the pillow basalts erupted in the oceanic ridge to form this tract of oceanic crust. The youngest fossils are the remains of animals that lived during Miocene time, perhaps as recently as 10 million or so years ago. Evidently part of this region was still under water that recently.

Hoisting the Ocean Floor

It is distinctly abnormal to find a large area of oceanic crust above sea level with hills and valleys eroded in it, and people living on it. What happened? The answer to that question lies north of the Willapa Hills, in the Olympic Mountains. That range is a trench filling, a contorted mass of oceanic sediments that were scraped off the surface of sinking oceanic crust, and stuffed into a trench. The slab of relatively undeformed oceanic crust that lies flat beneath the Willapa Hills tilts steeply at the south end of the Olympic Range, and wraps around its eastern and northern sides in a great looping fold.

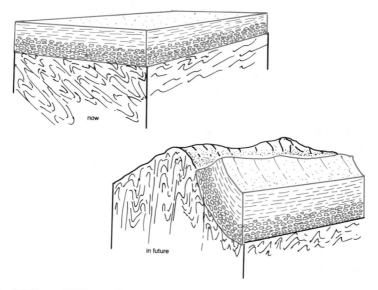

The Willapa Hills as they are now, and as they will be after the trench offshore ceases to exist.

We think the trench filling exposed in the Olympic Range continues south beneath the essentially horizontal slab of oceanic crust that is the Willapa Hills. Lithosphere sinking through the offshore trench stuffed light sedimentary rocks beneath the slab of oceanic crust in the Willapa Hills like small boy stuffing sticks of wood beneath a raft to make it float higher. The trench filling beneath jacked the Willapa Hills above sea level, and now lies waiting to become a new mountain range.

When the oceanic ridge now off the coast eventually arrives at the trench, about 10 or 15 million years from now, the trench and ridge will destroy each other. Then the San Andreas fault of California and Queen Charlotte Islands fault of British Columbia will meet to form a continuous transform plate boundary along the entire west coast from Mexico to Alaska. When oceanic crust no longer sinks through the trench, nothing will hold the trench filling down, and it will be free to float.

Then the rising trench fillings, a southward extension of the Olympic Range, will tilt the slab of oceanic crust beneath the Willapa Hills steeply eastward. Eventually, mountains composed of the rocks stuffed into that trench will extend down the coast to meet the Coast Range of California, which formed in precisely the same way when the trench and ridge met along that stretch of coast sometime between 25 and 15 million years ago. Like the future Coast Range of Washington, the California Coast Range has a steeply tilted slab of oceanic crust lying along its eastern flank.

Pillow Basalts

The oldest rocks in the Willapa Hills are the pillow basalts. Watch for rather evil looking black or greenish-black rock full of "pillows" about the size of barrels. Some are no larger than nail kegs. If you get a chance to see basalt pillows in three dimensions, you will find that they are actually long cylinders. The upper part of the oceanic crust, which covers about two-thirds of the earth's surface, consists mostly of basalt lava flows including pillow basalts.

Oceanic Sediments

As the newly formed oceanic crust moves away from the crest of the ridge where the basalt erupts, it picks up a cover of sediments. They consist mostly of mud and muddy sand eroded from the continent, and carried to the coast in streams. Waves spread the sediment across the continental shelf offshore. There, occasional disturbances, such as heavy storm waves, underwater landslides, or earthquakes, stir large deposits of mud into suspension to form clouds of muddy water.

Water that contains suspended mud is denser than clear water, so it pours down the continental slope, and then billows out across the deep ocean floor. As the mud finally settles to form a layer of sediment, the denser particles sink faster and the finer fraction slower. Therefore, layers of sediment on the deep ocean floor contain sand at the base and grade up into clay at the top. Geologists call such sediments turbidites, because clouds of turbid water deposit them.

Geologists recognize several formations of turbidite sedimentary rock on top of the pillow basalt in the Willapa Hills. They look very much alike. All the turbidite formations in the Willapa Hills contain layered gray, dark-gray, or greenish gray muddy sandstones and mudstones. Most of the rocks are fairly soft, so they rarely form good outcrops except in road cuts.

Oil and Gas

Petroleum typically occurs in sedimentary rocks laid down in sea water, so it seems reasonable to suspect that it may exist in the Willapa Hills. There is a gas field, the Mist Field, in northwestern Oregon, which produces from rocks identical to those in the southwestern part of the Willapa Hills. So it will be no surprise if someone discovers gas nearby in Washington.

The most promising part of the Willapa Hills for oil and gas prospecting is probably in the area between Grays Harbor and Chehalis. Sedimentary rocks are thickest in that part of the Willapa Hills, so a well there has a greater chance of penetrating a productive layer. The most basic strategy in exploration is to drill along the crests of anticlinal arches folded into the sedimentary rocks, because that is where any oil or gas that may exist will most readily accumulate. So far, none of the almost 50 wildcat wells drilled between Grays Harbor and Chehalis have produced petroleum in commercial quantity. However, many have found encouraging shows of oil and gas, and the drilling continues.

Flood Basalts

Several of the flood basalt lava flows that built the Columbia Plateau poured down the Miocene valley of the Columbia River to the Pacific Ocean. Some of those flows overflowed the river valley, and spread over large areas of southwestern Washington and western Oregon. Erosion has since removed most of that Miocene basalt, but large patches remain in the southwestern part of the Willapa Hills.

The Landscape

Except in the area south of Olympia, where glacial ice covered the northern end of the Willapa Hills and dumped sediment into the streams, most of southwestern Washington has a relatively simple landscape.

Throughout the Willapa Hills we see a gently rolling landscape of smoothly rounded hills with summits convex to the sky and fairly straight slopes that curve gently downward into the valley floors. Geologists call that a ridge and ravine landscape. It is the typical countryside of humid regions that support a dense plant cover. The landscape owes its character to the plants that protect the soil from most processes of erosion, except landsliding and soil creep.

Landslides are fairly common in the Willapa Hills simply because the region has a deep cover of wet soil that rests on weak bedrock. Most landslides are sufficiently obvious as large masses of moving hillside that disrupt roads and fences, and destroy buildings. Anyone thinking of building on a steep slope in the Willapa Hills should beware.

Soil creep is much more pervasive, but does not pose a threat. It is an extremely slow process that moves soil down the slope at a rate that rarely exceeds one eighth of an inch per year. Causes of soil creep include expansion and contraction of the soil through freezing and thawing, or wetting and drying, as well as the formation and subsequent collapse of holes created by roots and burrowing animals. In the Willapa Hills where the climate is too mild to cause much freezing and too wet to permit much drying, burrowing worms and plant roots probably cause most of the erosion on hillslopes.

Although the process is extremely slow, we can often see the effects of soil creep because it moves the surface levels of the soil much faster than those at depth. That makes fence posts pitch down slope, and retaining walls fail from the top down. Trees continue to grow upward as soil creep rotates their trunks, so they acquire a downslope slouch when they grow on creeping slopes.

Grays Harbor and Willapa Bay

During the last ice age, so much water was tied up in the great continental glaciers that sea level dropped about 300 feet. The beach was several miles offshore then, where the water is now 300 feet deep, and the area that is now the coast was several miles inland. When the last ice age ended, about 10,000 years ago in round numbers, the glaciers melted rather rapidly, and sea level reached its present stand sometime around 8000 years ago. The exact time is a matter of some dispute.

As the last of the great ice sheets melted, the ocean flooded into river valleys along the coast to make great bays, their mouths open to the sea. Then the waves eroded the projecting headlands that separated those bays, and washed their debris along the coast to build the long sand spits that now nearly bar the entrances to Grays Harbor and Willapa Bay. Both of those big bays now collect sediment that the rivers bring in from the land and the waves wash in from the sea. Before too many thousands of years have passed, both will become dry land.

I-5
Seattle—Portland
178 mi. 286 km.

Between Seattle and Olympia, the highway follows the east side of
Puget Sound across countryside so deeply underlain by glacial sediments that no bedrock is visible. The glacial deposits include till
dumped directly from the ice and layered outwash deposited from
meltwater flowing off the ice. In the southern part of the Puget Sound
lowland, they also include thick layers of sticky gray clay deposited in
a lake that flooded that area when the ice blocked the normal drainage through the Strait of Juan de Fuca. The fill of glacial debris is
about 3000 feet deep beneath Seattle, 2000 feet at Tacoma. Bedrock
finally appears near Olympia.

Between Olympia and Vancouver, I-5 follows an almost directly
north-south route across the Willapa Hills. Soils are deep in this
gently rolling countryside, and the vegetation dense, so roadcuts
provide the best rock exposures.

Ocean Floor Basalt

The oldest of those rocks are basalts erupted on the floor of the
Pacific Ocean. Eocene fossils in the sedimentary rocks between and
on top of the lava flows show that they erupted about 50 million years
ago. Little basalt appears along the road, although it forms the
bedrock in large areas not far west of the highway. Fortunately, there
is a nice roadcut in basalt on the west side of the highway less than a
mile south of the junction with U.S. 101.

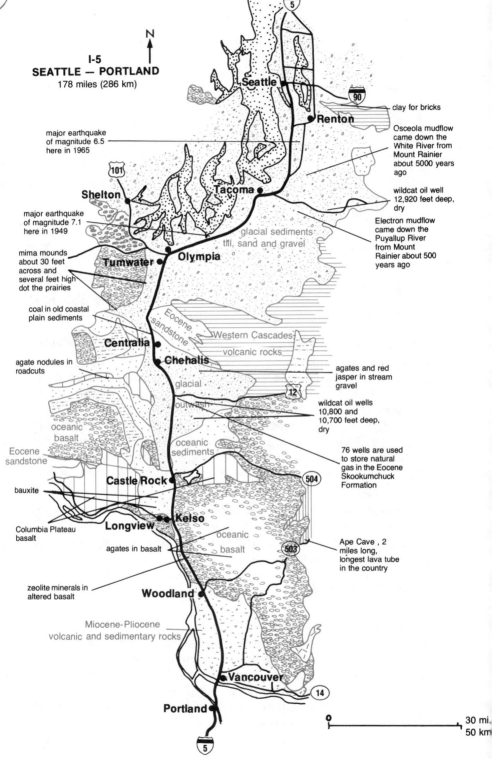

I-5
SEATTLE — PORTLAND
178 miles (286 km)

N

clay for bricks

Osceola mudflow came down the White River from Mount Rainier about 5000 years ago

major earthquake of magnitude 6.5 here in 1965

wildcat oil well 12,920 feet deep, dry

Electron mudflow came down the Puyallup River from Mount Rainier about 500 years ago

major earthquake of magnitude 7.1 here in 1949

glacial sediments: till, sand and gravel

mima mounds about 30 feet across and several feet high dot the prairies

coal in old coastal plain sediments

Eocene sandstone

Western Cascades volcanic rocks

agate nodules in roadcuts

glacial outwash

agates and red jasper in stream gravel

wildcat oil wells 10,800 and 10,700 feet deep, dry

oceanic basalt

Eocene sandstone

oceanic sediments

76 wells are used to store natural gas in the Eocene Skookumchuck Formation

bauxite

Columbia Plateau basalt

oceanic basalt

Ape Cave , 2 miles long, longest lava tube in the country

agates in basalt

zeolite minerals in altered basalt

Miocene-Pliocene volcanic and sedimentary rocks

Seattle
Renton
Shelton
Tacoma
Tumwater
Olympia
Centralia
Chehalis
Castle Rock
Longview
Kelso
Woodland
Vancouver
Portland

0   30 mi.
0   50 km

236

This roadcut near Kelso exposes a vertical dike of dark andesite cutting tan sandstone. The sandstone accumulated on the nearby floor of the Pacific Ocean some 50 million years ago. The andesite dike injected a fracture while the Western Cascades were active. Water penetrating fractures in the andesite weathered it into rounded pieces.

Oceanic Sediments

As we would expect in an uplifted slab of ocean floor, a thick section of sedimentary rocks, mostly dark gray, and greenish gray muddy sandstones lies above the basalt. Most of the roadcuts between Olympia and Vancouver expose those sediments, and they generally contain at least a few fossils.

The Old Coastal Plain

The interstate highway follows a route far enough east to take it close to the western edge of the North Cascade subcontinent, but the contact is buried beneath Western Cascade volcanic rocks and therefore impossible to locate exactly.

It is no surprise that a route so close to the North Cascade subcontinent crosses some sedimentary rocks that accumulated along its coast. The Skookumchuck formation consists of sedimentary rocks originally deposited along the west coast of the North Cascade subcontinent during Eocene time. The rocks vary considerably because they consist of both sedimentary material and volcanic ash that was deposited in shallow water, on land, and in tidewater marshes. The marsh deposits include layers of coal.

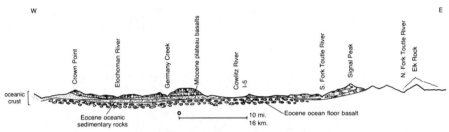

Section across the line of I-5 near Castle Rock.

Coal

The Centralia power plant, in the hills about 5 miles northeast of town, is the first large coal fired electric generating plant in the Pacific Northwest. It started operation in 1971, using coal produced from large open pit mines in the Skookumchuck formation nearby. There are large coal seams in the area, one of them as much as 50 feet thick, and several lesser deposits — enough to supply the plant for at least 35 years. Low transportation costs provide some solace for the poor quality of the coal.

Diversion of the Cowlitz River

In the area south of Chehalis, the highway passes through a low region of the Willapa Hills much too broad to be an eroded valley. It must be a block of the earth's crust dropped along faults. The Willamette Valley of Oregon contains such dropped blocks, and others may exist in the Puget Sound lowland. So it is not surprising to find one here, right in line.

Mount St. Helens

Washington 504 follows a route west from the interstate at Castle Rock up the Toutle River to Mount St. Helens. The route crosses old oceanic rocks along the eastern edge of the Willapa Hills, and then onto Cascade volcanic rocks.

Of course the big attractions are the mudflow in the Toutle River, the airborne ash that covers everything, and the ash flow near Mount St. Helens. Much of the ash and mudflow debris is already amazingly solid rock. Each particle of volcanic ash has many sharp edges that catch on each other and interlock to make a coherent deposit.

Immediately after the eruption, the floor of the Toutle River Valley and the area near Mount St. Helens were so desolate that restoration of the natural plant cover seemed an impossible dream. Nevertheless, plants started growing in much of the devastated area before the summer of 1980 ended. Since then, a steadily increasing variety of plants have covered more of the area every season, and will soon make it green again. Devastation like that wrought in the 1980 explosion of Mount St. Helens has happened many times in the Cascades.

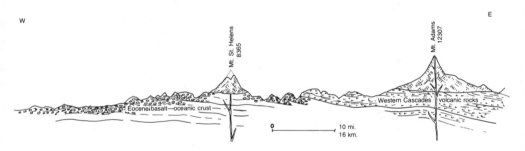

Section across Mount St. Helens.

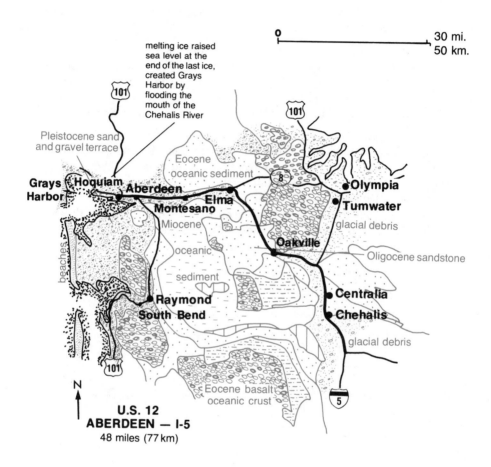

melting ice raised
sea level at the
end of the last ice,
created Grays
Harbor by
flooding the
mouth of the
Chehalis River

0   30 mi.
        50 km.

101

Pleistocene sand
and gravel terrace

Eocene
oceanic sediment

8

Grays **Hoquiam**
Harbor   **Aberdeen**   **Elma**   **•Olympia**
        Montesano   **Tumwater**

Miocene   glacial debris

oceanic   **Oakville**

sediment   Oligocene sandstone

beaches   **Raymond**   **Centralia**
        South Bend   **Chehalis**

                        glacial debris

101

N
↑

Eocene basalt
oceanic crust

5

U.S. 12
ABERDEEN — I-5
48 miles (77 km)

U.S. 12
Aberdeen — I-5
48 mi. 77 km.

U.S. 12 leaves the interstate near the community of Grand Mound, about 15 miles south of Olympia, and follows the Chehalis River west to Aberdeen at the head of Grays Harbor.

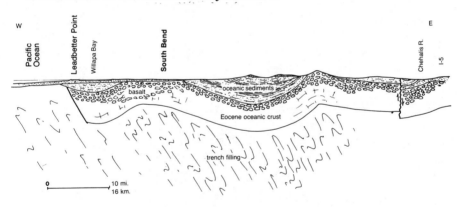

Section across the Willapa Hills along a line south of U.S. 12.

Old Ocean Floor

As everywhere in southwestern Washington, the older bedrock consists of basalts and pillow basalts erupted on the floor of the Pacific Ocean, and covered in most areas by sedimentary rocks that accumulated on the ocean floor. Nearly all the roadcuts between the interstate highway and Aberdeen expose those rocks. The basalts are dark green or greenish black, and make messy looking outcrops. The sedimentary rocks are gray or tan, and lie in neatly defined layers.

This highway crosses areas where the sedimentary cover on the basalt is deeper than in most parts of the Willapa Hills, and therefore more likely to contain oil and gas. At Montesano, the road crosses the Caldwell Creek anticline, an arch folded into the sedimentary layers. That is where any oil and gas that may exist in the neighborhood are probably trapped, and therefore the target of considerable wildcat drilling. So far, there is no commercial production, but some wells have produced enough gas to inspire optimism. Some day, there may be a gas field near Montesano.

Plateau Basalt

After the old ocean floor had risen to or sightly above sea level, flood basalt lava flows of the Columbia Plateau covered them in large areas near the mouth of the Columbia River. Although outcrops of flood basalt do exist near this road, none are visible from the highway.

Glacial Meltwater

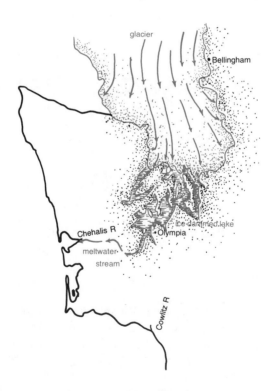

Glacial meltwater drainage during the last ice age.

The Chehalis River and its valley are the most striking geologic sights along this route. Notice that the river is much too small for its valley, so undersized that it seems lost on the broad expanse valley floor. Evidently, the modern river is a shrivelled remnant of a much larger stream that once flowed through this valley.

During the last ice age, and presumably during earlier ice ages, a large glacier pushed south out of Canada into the Puget Sound lowlands. As the ice reached about as far as Bremerton and Seattle, it blocked the Strait of Juan de Fuca, and thus dammed the streams draining that way. A lake formed in the area between the glacier to the north and the low hills surrounding the southern end of the Puget Sound lowland.

When the surface of that lake reached an elevation of about 200 feet, water poured over the lowest pass through the hills and formed a spillway that followed the valley of the Black River to where it joins the valley of the Chehalis River about midway between Rochester and Oakville. When that happened, the valley may have belonged to the Cowlitz River, but that is another story.

In any case, the old meltwater channel that the Black and Chehalis rivers now follow is about the size of the modern valley of the Columbia River. Therefore, it seems reasonable to conclude that flow through that valley must have been comparable to that of the Columbia today. Of course, that flow combined the discharge of all the streams that normally empty into the southern end of the Puget Sound lowlands. Even so, the flow must have been greater than all those streams could muster. Evidently, the climate was wetter during the last ice age than it is today.

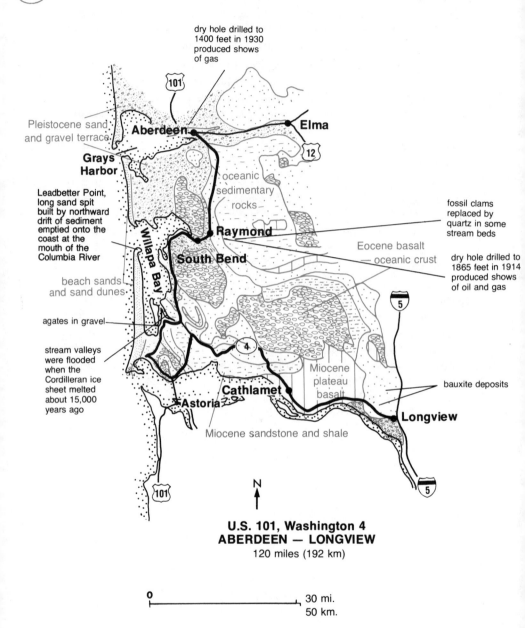

dry hole drilled to
1400 feet in 1930
produced shows
of gas

101

Aberdeen

Elma

Pleistocene sand
and gravel terrace

**Grays
Harbor**

12

oceanic
sedimentary
rocks

Leadbetter Point,
long sand spit
built by northward
drift of sediment
emptied onto the
coast at the
mouth of the
Columbia River

Raymond

fossil clams
replaced by
quartz in some
stream beds

South Bend

Eocene basalt
— oceanic crust

dry hole drilled to
1865 feet in 1914
produced shows
of oil and gas

beach sands
and sand dunes

Willapa Bay

5

agates in gravel

stream valleys
were flooded
when the
Cordilleran ice
sheet melted
about 15,000
years ago

4

Miocene
plateau
basalt

bauxite deposits

Cathlamet

Astoria

Longview

Miocene sandstone and shale

101

N
↑

**U.S. 101, Washington 4
ABERDEEN — LONGVIEW**
120 miles (192 km)

5

0   30 mi.
         50 km.

244

U.S. 101, Washington 4
Aberdeen — Longview
120 mi. 192 km.

Between Aberdeen and Raymond, the road follows a straight north and south path through the western part of the Willapa Hills. All the bedrock in this area formed on the floor of the Pacific Ocean. It consists of basalts and pillow basalts erupted in the crest of the Juan de Fuca ridge, and of dark gray and greenish gray sedimentary rocks deposited on the basalt. The sedimentary rocks are at the surface along the northern half of the route, the basalts along the southern half. Unfortunately, neither is especially well exposed in this region of deep soils and dense vegetation.

South of Raymond, the road follows the Willapa River to the head of Willapa Bay, and then follows the shore of that bay almost to Astoria. The bedrock, if we could only see it, is still the old floor of the Pacific Ocean. But the main geologic attraction here is the bay.

Willapa Bay

When sea level was 300 feet lower during the last ice age, Willapa Bay was high and dry, part of the Willapa Hills. Sea level rose to its present stand as the meltwater ran back to the ocean at the end of the ice age, and the Pacific surf pounded on what is now the inner shore of Willapa Bay. During the last 8000 or so years, the waves have built the baymouth bar that nearly isolates Willapa Bay from the ocean, and protects it from the surf. Washington 103 follows that bar to its northern tip at Leadbetter Point, about 20 miles of sandspit built in less than 10,000 years.

Willapa Bay seen from Leadbetter Point, at the end of Washington 103. The beach grass, a European immigrant, traps sand that blows inland from the low-tide mudflats.

Along this coast, the prevailing winds tend to blow from the southwest in winter, from the northwest in summer. The bar across the mouth of Willapa Bay extends north from the mouth of the Columbia River, so it seems reasonable to conclude that it consists mostly of sediment dumped at the mouth of the river, and then moved north during winter storms. Now that so many dams impound sediment coming down the river before it can reach the coast, that sandspit will grow much less rapidly.

Beach grass partially stabilizes these dunes near Leadbetter Point.

The internal anatomy of a sand dune as exposed by a spade. The steeply dipping layers form as the wind blows sand over the sharp edge of an advancing dune, and dumps it on the steep downwind slope.

At low tide, the wind whips sand off the beach and mudflats, and blows it inland into dunes, which build the sandspit higher above sea level. So a visit to the beach is also a visit to a sea of sand dunes. The beaches of North America contain far more wind blown sand than its deserts.

Astoria — Longview

Washington 401 winds its way north from the Astoria bridge to Washington 4, which winds along the north side of the Columbia River to Longview. Between Cathlamet and the western outskirts of Longview, the road follows the base of high bluffs eroded in flood basalt lava flows of the Columbia Plateau. Those incredible flows really did go all the way from easternmost Washington and Oregon to the Pacific Ocean.

The modern landscape in this area consists of flat-topped hills, the original flow surface, separated by modern stream valleys eroded into that surface. Red laterite soil that formed during Miocene time, when the climate was wet and tropically warm, covers those flat uplands, and drapes down the walls of the valleys that separate them. Much of that laterite is the aluminum-rich variety called bauxite.

Bauxite

All lateritic soils consist of iron oxide, aluminum oxide and clay in various proportions. Near the coast, some of the lateritic soils developed on basalt lava flows became especially rich in aluminum oxide, and poor in the other constituents. That kind of laterite soil, called bauxite, provides the world's only important source of aluminum ore.

Bauxites in this part of western Washington are about 12 feet thick, and full of hard lumps of white aluminum oxide, some as big as small potatoes. The soils are red or yellow because they contain iron oxide, which must be removed before the bauxite is reduce to metallic aluminum. Washington bauxite has never been mined, except experimentally, because it contains too much iron to compete commercially with aluminum ore from the Caribbean region, and northern South America.

VII
THE OLYMPIC PENINSULA AND PUGET SOUND LOWLAND

The geologic map shows the slab of oceanic crust that lies flat beneath the Willapa Hills wrapped around the Olympic Peninsula in a great fold shaped like a fancy horseshoe lying with its open end facing west. It holds in its embrace, the Olympic Mountains in the core of the peninsula. The high mountains are a mass of chaotically deformed muddy sandstone, oceanic turbidites similar to those in the slab of oceanic crust wrapped around them except in being severely deformed. They even contain Eocene fossils, and must therefore have been laid down on the basalt at the edge of the ocean floor at the same time.

However, the sandstones in the core of the Olympic fold lie beneath the pillow basalt that wraps around the peninsula. That pillow basalt .is the bedrock of the oceanic crust, so the sandstone could not have been deposited beneath it, but must have gotten there in some other way. That is the major geologic problem of the Olympic Peninsula.

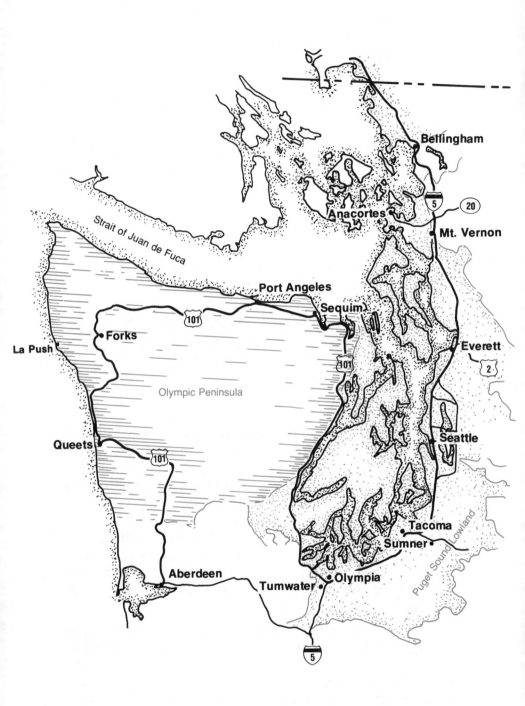

The Olympic Peninsula and Puget Sound lowland.

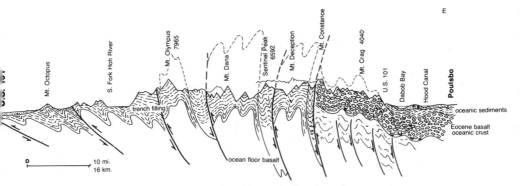

Section across the Olympic Peninsula.

Geologists who study the sandstone in the Olympic Mountains find it most severely deformed in their inner core, progressively less so closer to the coast. Those patterns, and the generally chaotic style of the deformation, are typical of oceanic sediments that were stuffed into an oceanic trench. Evidently, the slab of ocean floor sinking off the west coast, stuffed that crushed and broken sandstone under the tract of oceanic crust that now wraps around the Olympic Peninsula. The core of the Olympic fold is a sample of what exists beneath the Willapa Hills in southwestern Washington and the Coast Range of Oregon.

THE PUGET SOUND LOWLAND

The Puget Sound lowland is not a geologic province in the ordinary sense of the idea. It is simply a region so deeply buried beneath glacial deposits that little can be seen or said of the bedrock. In treating the Puget Sound lowland as a separate region, we surrender to the opacity of glacial sediments as much as 3000 feet thick. The obvious presumption that bedrock beneath the glacial debris in the Puget Sound lowland must resemble that in the nearby Olympic Peninsula, Willapa Hills and North Cascades is reasonable, but not completely satisfying. Why the lowland?

Although glacial debris floods the Puget Sound area, we can not reasonably attribute the existence of the lowland to glacial

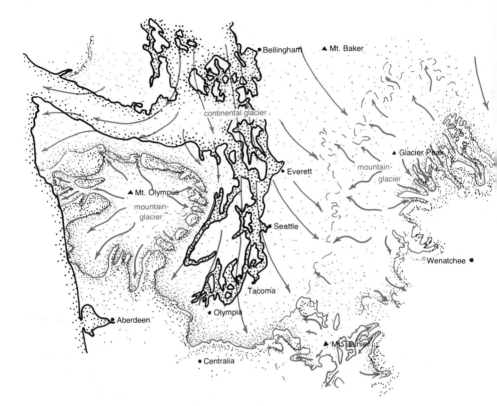

Approximate maximum extent of glaciation in the Olympic Peninsula and Puget Sound lowland.

erosion. Those glaciers were, after all, near the southern limit of their reach, where the thinning ice was far more inclined to deposit sediment than to erode bedrock. The glaciers that filled the Puget Sound lowland could not have gouged it out.

The Puget Sound lowland lies directly in line with the Willamette Valley of western Oregon, which appears to contain a series of bedrock basins dropped along faults and filled with Pliocene gravel. A similar gravel-filled basin exists in the Willapa Hills south of Chehalis. It seems reasonable to project that trend north, and suggest that the Puget Sound lowland exists because a slice of the earth's crust dropped along faults. Fault movement forming basins in a trend parallel to the coast may well relate to the continuing northward drag of the western margin of North America.

OLYMPIC NATIONAL PARK

The park consists of two parts: a large area of high mountains in the interior of the Olympic Peninsula, and a narrow strip along the beach from Queets north to Cape Alava, almost at the tip of the peninsula. Neither part is more than marginally accessible by road. This is mostly a hiker's park.

The Glaciated Mountains

The closest road access to the mountainous interior is by way of Hurricane Ridge, on the north side of the mountains. The road turns south from U.S. 101 a few miles west of Port Angeles, and then winds tortuously from sea level to an elevation of about 6000 feet. From the end of that road, hikers can follow good trails into a vast wilderness of glacially sculptured peaks and valleys. High glaciated peaks in the Olympic Mountains provide good habitat for geologists. Very few places provide such excellent exposures of sedimentary rocks that were deposited along the edge of the ocean floor and then crammed into a trench. It is difficult to see as much at lower elevations where deep soil and a lush growth of plants cover the rocks.

In a sense, a sort of ice age still continues in the Olympic Mountains, which contain several of the largest and healthiest glaciers in the lower 48 states. And glaciers descend to much lower elevations in the Olympic Mountains than in the Cascade or Northern Rocky mountains even though the inland ranges are much colder. The existence of big glaciers in the Olympic Mountains where the climate is relatively mild, and the absence of comparable glaciers in the much colder Rocky Mountains, suggests that heavy snow is more important than cold weather in maintaining glaciers.

Impressive as they are, the modern glaciers are mere shadows of those that thrived here during the last ice age. Those giants descended the larger valleys almost to sea level, and joined the ice that filled the Puget Sound and Strait of Juan de Fuca. They left their record in the landscape of the Olympic Peninsula.

Mountain glaciers rank among nature's most effective rock eroders and earth movers. Rocks and fine sediment embedded in the sole of a moving glacier rasp and polish bedrock surfaces. And the ice freezes tight to the bedrock, and then plucks blocks loose as it moves on. Ice age glaciers gouged the valleys of the Olympic Peninsula into straight troughs with steep sides that provide long views to distant craggy peaks, often reflected in sparkling lakes. As they gouged the valleys wider, the big glaciers scraped the ridges separating them to narrow fins of rock that rise to converge in jagged peaks. Moraines clearly preserve the outlines of the lower ends of the ice age glaciers. Outwash deposited from glacial meltwater fills the lower parts of the large valleys, and forms broad aprons covering areas of low country around the base of the mountains.

The Coast

The highway follows the beach portion of Olympic National Park between the south side of the Hoh River and Queets, a distance of about 12 miles. Numerous trails provide access from the road to that section of the beach. A side road reaches the beach at La Push, about 15 miles north of Queets. Only hikers visit the coast between Queets and La Push. Hiking

A wave cut bench exposed at low tide.

along the coast is mostly a low tide sport. Walk along the beach, where there is a beach, and follow the trail over the hills and through the woods where the waves break directly against the cliffs. Trails exist wherever they become desirable, generally to provide shortcuts across rocky headlands.

This is a steep coast where the waves pound directly on the rocks, and carve them into the characteristic shapes of the shore. Breaking waves erode the sea cliff partly by battering it with loose pebbles, mainly by pounding pulses of compressed air into fractures in the rock. Each wave compresses the air in the fractures, thus gradually working blocks of rock loose, and popping them out into the surf. Waves can erode only to the high tide level, so they cut a notch in the base of the sea cliff until, undermined, it collapses. Then, the waves cut a new notch, and work the cliff back farther. Thus the coast encroaches into the land across a broad surface of wave cut bedrock, which geologists call the wave cut bench.

Because it exploits fractures so tirelessly, the surf quickly erodes masses of broken rock, opening them into channels, caves, and even tunnels. Meanwhile, the less fractured rock, which offers fewer opportunities to the waves, resists erosion. As the coast retreats, isolated stalwarts of resistant rock, called stacks, remain standing in the surf. Even they finally yield to

Stacks standing in the surf.

the relentless pounding. Waves focus the energy of their attack on projecting headlands, and erode them most rapidly. In many places, we now see an armada of stacks standing in the surf to mark where a ridge projected out to sea as a headland only a few thousand years ago. Meanwhile, the waves wash the debris they erode from the headlands into quiet coves, filling them with sediment. So, regardless of the original shape of a coast, the waves straighten the coastline by eroding the headlands and filling the coves.

Pockmarked Cliffs

Look closely at the pattern of shallow pockets eroded into the dark gray turbidite sandstone exposed in the sea cliffs, especially on surfaces above the limit of the waves. Such pitted surfaces are rare in places beyond the reach of salt spray. Geologists believe they form through a process called salt weathering, in which salt crystallizing in the rock wedges the mineral grains apart.

The Coastal Landscape

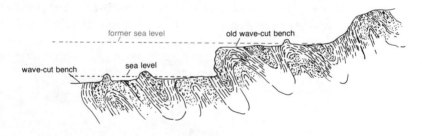

Cross-section of a coast showing the modern wave cut bench, and an older one now raised above sea level by uplift of the land.

Long stretches of coastal highway follow smoothly graded bench surfaces above the sea cliff. In some areas, it is possible to recognize several such benches, one above the other, rising landward from the coast like a flight of giant steps. Those coastal benches resemble in every way the wave cut benches that extend seaward from the base of the modern sea cliffs, and appear to be old wave cut benches, now above sea level. Some still have old sea stacks still standing on them.

256

It is difficult to interpret those old wave eroded surfaces as a record of series of drops in sea level, because similar terraces do not occur at the same elevations on other coasts. Instead, they seem to record a series of more local uplifts of the land. That could easily happen as the sinking slab of oceanic crust jams more slices of relatively light sedimentary rock into the trench, raising the rocks at the surface.

The Rocks in the Sea Cliffs

Boring clams drilled these holes into dark gray oceanic sandstones exposed just below high tide level.

All the rocks exposed in the sea cliffs accumulated as sediments on the floor of the Pacific Ocean. Most consist of rather fine-grained dark gray sandstones. However, there are occasional exposures of conglomerate, a rock consisting of nicely rounded pebbles, now stuck tightly together. In several places between Queets and La Push, the sea cliffs expose a rock called breccia that consists of large and small angular fragments of broken sandstone cemented together into solid rock. People call them "smell rock" because freshly broken pieces reek of crude oil.

It appears that masses of broken rock pushed up through the trench filling to make large intrusions of breccia. Evidently, oil rises into those intrusions and permeates the rock, filling the spaces between broken fragments. In several places, oil actually seeps from the ground. More than a dozen wildcat wells have been drilled along this coast. Most found good shows of oil, and one actually produced oil for a short time. Development of commercial oil production near the beach portion of Olympic National Park is probably only a matter of time.

I-5
Seattle — Vancouver, B.C.
141 mi. 225 km.

The drive between Seattle and Vancouver could as well be included in the section dealing with the North Cascade subcontinent, because that is where all the bedrock along the way belongs. However, except for the short stretch between Sedro Woolley and Bellingham, the road crosses glacial debris almost exclusively, so wave at the hills of older rock in the distance as you cross deposits of glacial till and outwash.

Glaciation

Glaciers filled the Puget Sound lowland at least several times during the ice ages, and left the entire region heavily plastered with glacial deposits, which express themselves primarily in the landscape. Notice the crude north to south grain of topography in the tendency of ridges, lakes, and arms of Puget Sound to stretch from north to south. That reflects the general southward flows of the ice. Unfortunately, perching a major metropolitan area on the glacial landscape makes it difficult to see more than the largest and most obvious of the original landforms.

Landslides are one troublesome result of building a city on glacial debris. In nearly all cases, excess of water in the ground actually triggers landslide movement. Dry ground rarely slides. Unfortunately, urban development commonly does cause water to enter the ground through such sources as drain fields, and leaking water mains, storm drains or sewers. Even so, landslides also tend to develop in places where the rocks naturally favor them. Those in the Seattle metropolitan area typically form in one of two geologic situations.

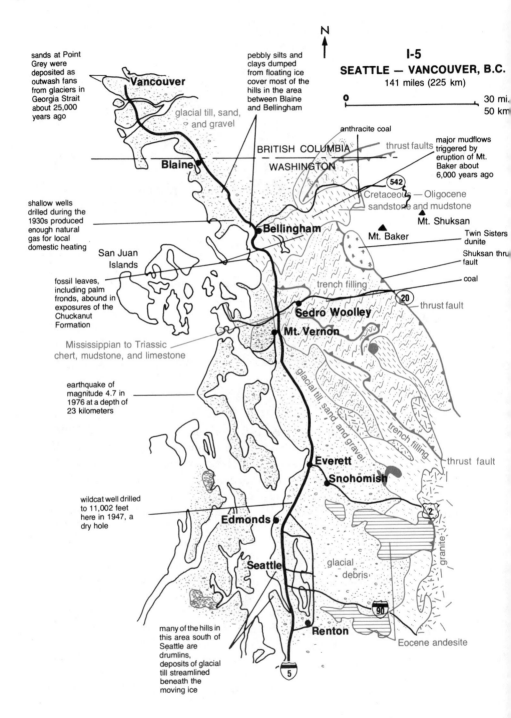

sands at Point Grey were deposited as outwash fans from glaciers in Georgia Strait about 25,000 years ago

Vancouver

glacial till, sand, and gravel

pebbly silts and clays dumped from floating ice cover most of the hills in the area between Blaine and Bellingham

N

I-5
SEATTLE — VANCOUVER, B.C.
141 miles (225 km)

0 .... 30 mi.
.... 50 km

anthracite coal

thrust faults

major mudflows triggered by eruption of Mt. Baker about 6,000 years ago

BRITISH COLUMBIA
WASHINGTON

Blaine

Cretaceous — Oligocene sandstone and mudstone

▲ Mt. Shuksan

shallow wells drilled during the 1930s produced enough natural gas for local domestic heating

Bellingham

▲ Mt. Baker

Twin Sisters dunite

Shuksan thrust fault

San Juan Islands

fossil leaves, including palm fronds, abound in exposures of the Chuckanut Formation

trench filling

coal

20

thrust fault

Sedro Woolley

Mississippian to Triassic chert, mudstone, and limestone

Mt. Vernon

earthquake of magnitude 4.7 in 1976 at a depth of 23 kilometers

glacial till, sand and gravel

trench filling

thrust fault

Everett

Snohomish

wildcat well drilled to 11,002 feet here in 1947, a dry hole

2

Edmonds

granite

Seattle

glacial debris

90

many of the hills in this area south of Seattle are drumlins, deposits of glacial till streamlined beneath the moving ice

Renton

Eocene andesite

5

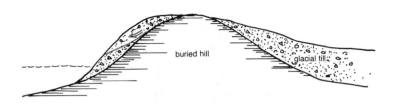

Landslides often develop in places where glacial debris slips off a hill of older rock.

In some areas, young and unconsolidated glacial debris thinly plasters hills of older and more solid rock — either bedrock or glacial deposits older than the last ice age. If such a slope gets too wet, the glacial debris may simply slide off, exposing the older material beneath.

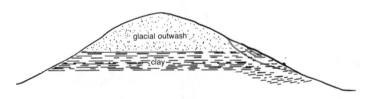

Landslides are likely where glacial outwash lies on top of a bed of clay.

Everyone has at some time slipped in the mud, and therefore knows from personal experience that clay gets slippery when wet. That is just as true inside hills as on a forest trail in the rain. Furthermore, beds of clay tend naturally to get wet because water can not drain through them. Where sand and gravel deposits of glacial outwash lie on a bed of clay, water soaking down through the ground will accumulate on top of the impermeable clay, making it extremely wet, weak, and slippery. The mechanical effect is as though the sand and gravel were lying on a bed of grease. Eventually, everything above the clay slips off in a landslide, which is destructive if it moves in a developed area.

The Western Cascades, and the Old Coastal Plain

Hills directly east of Seattle are mostly andesitic volcanic rocks erupted from the Western Cascade volcanoes. They probably lie on top of tightly folded rocks that belong to the old coastal plain of the

261

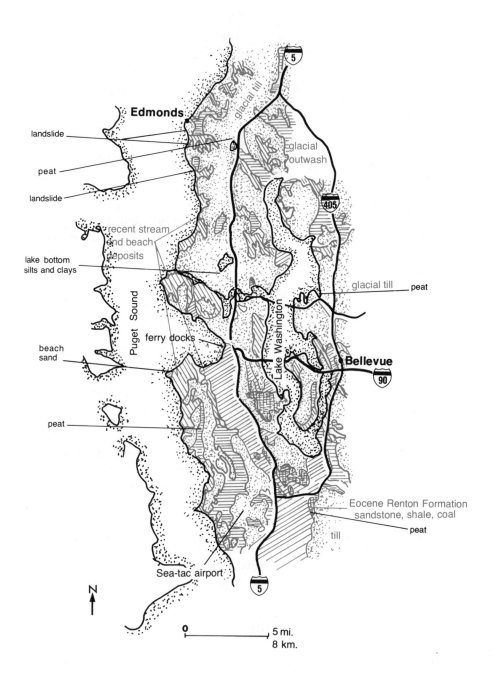

Geologic map of the Seattle area.

North Cascade subcontinent. Those rocks appear at the surface in the area southeast of Seattle, between Issaquah and Enumclaw.

The Seattle coal field, centered in the area around Black Diamond, southeast of Renton, began production in 1883, and peaked with a yearly output of about 900,000 tons shortly after the turn of the century. Then California oil began to capture the market, and coal production in the Seattle area dwindled steadily for decades, finally almost to nothing. Black Diamond and nearby communities that began as coal mining towns transformed themselves into remote bedroom suburbs of Seattle and Tacoma. Now that oil is no longer cheap, the Seattle coal field is resuming production, and the old coal mining towns are likely to resume their original identities.

The North Cascade Subcontinent

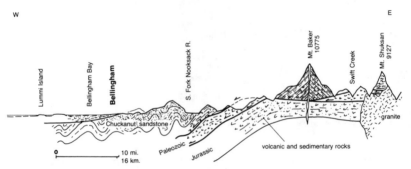

Section across the highway at Bellingham showing the older rocks of the North Cascade subcontinent on top of the Shuksan thrust fault, which separates them from younger oceanic crust beneath.

From Everett north, hills on both sides of the highway, including the San Juan Islands, belong to the North Cascade subcontinent. All consist of oceanic material, mudstone and basalt oceanic crust, that was stuffed into the trench along the west side of the North Cascade micro-continent before it joined North America.

Somewhere near Sedro Woolley, the exact spot hidden beneath glacial debris, the highway crosses the Shuksan thrust fault, which brings oceanic crust beneath the older rocks of the North Cascade subcontinent. That movement probably happened as the docking subcontinent telescoped the rocks along its margins. South of Sedro Woolley, the rocks buried under the glacial drift are probably beneath the Shuksan thrust fault, the same slab of oceanic crust that wraps around the eastern side of the Olympic Peninsula. From Sedro Woolley north, rocks beneath the road are on top of the Shuksan thrust fault, and belong to the North Cascade subcontinent.

A roadcut in the black Darrington phyllite, a deformed oceanic mudstone, beside the highway north of Sedro Woolley. The crumpling may reflect movement on the Shuksan thrust fault, which is not far below the surface in this area.

Mount Shuksan as seen from the road to Mount Baker. It is a mass of metamorphic rock resting on the Shuksan thrust fault.

Columns of Mount Baker andesite stand like long sticks of cordwood stacked on end in this outcrop near Mount Baker lodge.

Twin Sisters Dunite.

Twin Sisters Mountain, 24 miles west of Bellingham and almost immediately southwest of Mount Baker, is one of the most extraordinary bodies of rock in the country. The mountain, which is about 10 miles long and 3 miles wide, is a solid mass of an extremely rare rock called dunite, which consists almost entirely of the mineral olivine. Dunite properly belongs in the mantle, and is greatly out of place in a continent. The fresh rock is green, the color of olivine, but it weathers to the rusty brown color that makes the ragged top of Twin Sisters Mountain conspicuous from a great distance.

Dunite has the generally massive look of an igneous rock, but its melting point is so high that it could not have been molten. Perhaps it forms through metamorphic recrystallization of serpentinite, which has the right chemical composition, and could turn into a solid mass of olivine if it were strongly heated. It seems a shame that dunite is not terribly valuable. It does contain a small percentage of chromite, a black mineral valuable as a source of chromium, but not enough to make it worth mining. There is a limited market for the olivine for use in making fire brick, and as foundry sand, but that supports only a very small mining operation.

Mount Baker

Mount Baker appears on the eastern horizon from long stretches of the interstate highway. Next to Mount Rainier, it is the most accessible of the big volcanoes in Washington, and well worth the side trip from Bellingham on Washington 542. The road follows the North Fork of the Noocksack River, crossing mudflow deposits from Mount Baker much of the way.

U.S. 101
Olympia — Port Townsend
100 mi. 160 km.

Glacial deposits blanket the lowlands between Olympia and Potlatch so deeply that no bedrock appears at the surface. It seems safe to assume that the buried bedrock in this region is old oceanic crust similar to that exposed a few miles farther north and farther west. Between Potlatch and Port Townsend, the road does pass bedrock hills, all eroded in the tilted slab of oceanic crust that wraps around the Olympic Peninsula.

Hood Canal

Between Potlatch and Quilcene, the highway follows the west bank of the Hood Canal. The name is misleading because Hood Canal is actually a natural valley eroded along the trend of the bedrock. Hills west of Hood Canal consist mostly of basalts and pillow basalts once on the old ocean floor, those on its east side of sedimentary rocks deposited on the pillow basalts. Evidently, the original stream eroded its valley in the softer sedimentary rocks, and the glacier followed its example, no doubt guiding on the hard basalts along the west side of the valley. As a glaciated valley now flooded with sea water, Hood Canal is technically a fjord. Glacial deposits blanket the sides of the hills above Hood Canal to an elevation of about 1200 feet, so the ice was about that deep.

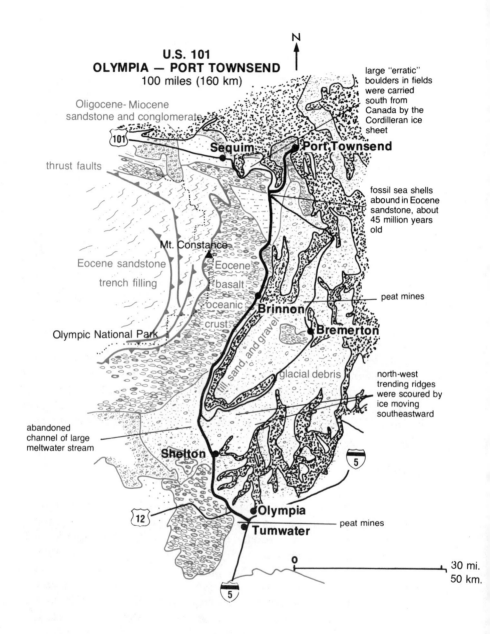

**U.S. 101
OLYMPIA — PORT TOWNSEND**
100 miles (160 km)

N

Oligocene- Miocene
sandstone and conglomerate

large "erratic"
boulders in fields
were carried
south from
Canada by the
Cordilleran ice
sheet

Sequim

Port Townsend

thrust faults

fossil sea shells
abound in Eocene
sandstone, about
45 million years
old

Mt. Constance

Eocene sandstone

Eocene
basalt

trench filling

oceanic

peat mines

crust

Brinnon

Olympic National Park

Bremerton

north-west
trending ridges
were scoured by
ice moving
southeastward

till, sand and gravel

glacial debris

abandoned
channel of large
meltwater stream

Shelton

5

12

Olympia

peat mines

Tumwater

0   30 mi.
            50 km.

5

268

U.S. 101
Port Townsend—Forks
94 mi. 152 km.

Between Port Townsend and Forks, the road follows the north edge of the Olympic Peninsula. All of the bedrock is part of the steeply tilted slab of oceanic crust that wraps around the Olympic Peninsula, the same rocks that lie flat beneath the Willapa Hills. However, glacial debris buries the bedrock along most of the route.

Glacial Deposits

During the last ice age, an enormous glacier filled the Strait of Juan de Fuca all the way to the ocean. The ice flowed south from British Columbia carrying distinctive Canadian rocks that now litter the north slope of the Olympic Mountains to an elevation of 4500 feet. Therefore the glacier in the Strait of Juan de Fuca must have been more than 4500 thick, considering that it extended to the bottom of the strait. Remember that sea level was about 300 feet lower at the maximum of the last ice age than it is today. Much smaller, but nevertheless very large, glaciers flowed down the high valleys of the Olympic Mountains to join the main glacier in the Strait of Juan de Fuca.

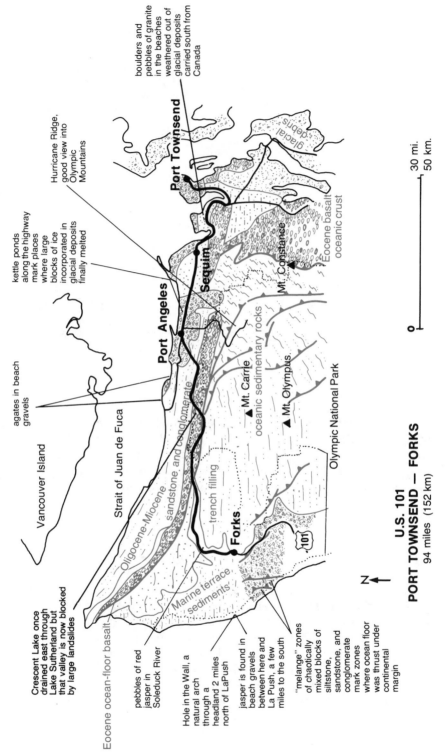

boulders and pebbles of granite in the beaches weathered out of glacial deposits carried south from Canada

Hurricane Ridge, good view into Olympic Mountains

kettle ponds along the highway mark places where large blocks of ice incorporated in glacial deposits finally melted

Port Townsend

glacial debris

Eocene basalt oceanic crust

Sequim

Mt. Constance

Port Angeles

agates in beach gravels

oceanic sedimentary rocks

Mt. Carrie

Mt. Olympus

Vancouver Island

Strait of Juan de Fuca

Oligocene-Miocene sandstone and conglomerate

trench filling

Olympic National Park

Crescent Lake once drained east through Lake Sutherland but that valley is now blocked by large landslides

Eocene ocean-floor basalt

Forks

Marine terrace sediments

101

pebbles of red jasper in Soleduck River

Hole in the Wall, a natural arch through a headland 2 miles north of LaPush

jasper is found in beach gravels between here and La Push, a few miles to the south

"melange" zones of chaotically mixed blocks of siltstone, sandstone, and conglomerate mark zones where ocean floor was thrust under continental margin

N

0   30 mi.
   50 km.

U.S. 101
PORT TOWNSEND — FORKS
94 miles (152 km)

The signs of large scale ice age glaciation abound along the north side of the Olympic Peninsula. Numerous ponds and bogs flood low places in the deposits of glacial debris that the glacier left on most of the countryside. And erratic boulders litter the landscape, still lying where the glacier left them as it melted at the end of the ice age. An unsuccessful wildcat oil well drilled near Sequim in 1956 passed through some 2100 feet of glacial deposits before it penetrated bedrock.

U.S. 101 skirts into the edge of Olympic National Park as it passes along the south shore of Crescent Lake, one of the most beautiful lakes in a state that has many. Crescent Lake floods a valley the glacier scoured deep along the trend of the rocks. Ordinarily, that valley would drain east into the Elwha River, and thence into the Strait of Juan de Fuca. However, large landslides blocked the drainage after the ice melted, impounding the lake. Much smaller Lake Sutherland, south of the road just east of Crescent Lake, has a similar origin.

Ozette Lake, just south of the western tip of the Olympic Peninsula looks on the road map like a coastal lagoon impounded behind a beach ridge. In fact, it is fresh water impounded behind a glacial moraine, a low boulder-strewn ridge about a mile wide that wraps around the seaward side of the lake. Of course, the coast was farther west when the moraine formed, because sea level was lower then.

Oceanic Crust and Turbidite Sediments

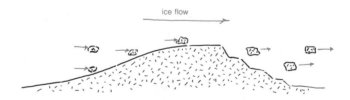

Moving glaciers abrade the upflow sides of bedrock knobs and pluck their downflow sides to create a steep and ragged slope that faces in the direction of ice movement.

The hills along the highway are glacially sculptured bedrock, and they express the westward flow of the ice in their shapes — steep sides facing generally west, in the downflow direction. Bedrock outcrops in those hills consist of oceanic rocks, of pillow basalts and the gray turbidite sandstones and mudstones that accumulated on them. Similar rocks appear in the sea cliffs. Most of the few bedrock expo-

271

A cliff exposure of basalt pillows about two feet in average diameter.

sures along the road on the north side of the Olympic Peninsula are dark greenish black or black pillow basalts, once the bedrock floor of the Pacific Ocean. Dark gray sandstone and mudstone deposited on top of the pillow basalt appears in occasional roadcuts, more abundantly in natural exposures in the sea cliffs.

Many outcrops of sandstone contain abundant fossils. The oldest are the remains of animals that lived during Eocene time, about 50 million years ago. Evidently the basalts beneath are about the same age. The youngest sedimentary layers contain fossils of animals that lived in relatively shallow sea water during Miocene time, as recently as 10 million years ago. Clearly, the region emerged to become dry land since then.

Agate Bay

Agate Bay, on the shores of the Strait of Juan de Fuca west of Port Angeles, is an excellent place to see pillow basalt exposed in the sea cliffs. The bay gets its name from the abundant beach agates that weather out of the basalt. Look for them when the tide is out.

Wave ripples exposed at low tide on the mudflats near Agate Beach.

U.S. 101
Forks—Aberdeen
116 mi. 187 km.

Most of the route between Forks and Aberdeen crosses rather flat terrain between the mountains and the ocean, most of it across old shallow sea floor now above sea level, the rest across broad fans of glacial outwash. Very little bedrock is exposed except in the sea cliffs below the short stretch of road that follows the coast between Ruby Beach and Queets.

Oil

Wildcat oil wells were drilled near the mouth of the Hoh River in 1913, and again during the 1930's, when a small community of tar paper shacks called "Oil City" briefly thrived. The drillers found numerous shows of petroleum, and several wells actually produced

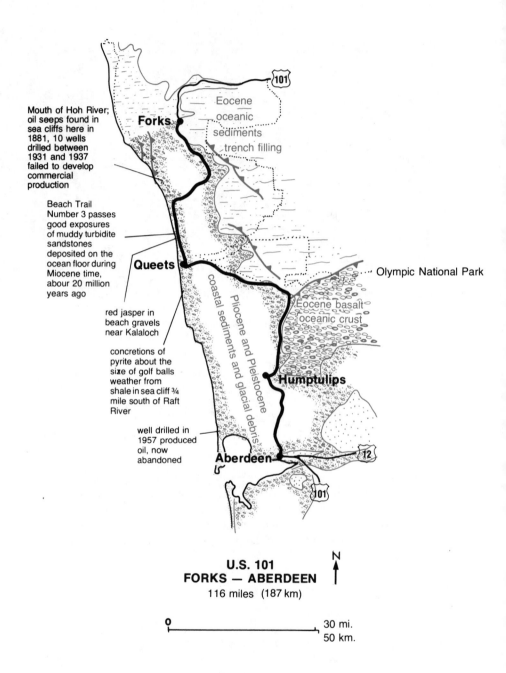

Mouth of Hoh River; oil seeps found in sea cliffs here in 1881, 10 wells drilled between 1931 and 1937 failed to develop commercial production

Forks

Eocene oceanic sediments trench filling

Beach Trail Number 3 passes good exposures of muddy turbidite sandstones deposited on the ocean floor during Miocene time, abour 20 million years ago

Queets

Olympic National Park

Eocene basalt oceanic crust

Pliocene and Pleistocene coastal sediments and glacial debris

red jasper in beach gravels near Kalaloch

concretions of pyrite about the size of golf balls weather from shale in sea cliff ¾ mile south of Raft River

Humptulips

well drilled in 1957 produced oil, now abandoned

Aberdeen

12

101

N

U.S. 101
FORKS — ABERDEEN
116 miles (187 km)

0   30 mi.
50 km.

A sea cliff near Kalaloch. The steeply tilted layers of sandstone in the lower part of the picture are deformed sediments deposited on the edge of the ocean floor. The upper surface of those sandstone beds is an old wave cut bench now buried beneath horizontal layers of wave deposited sand. Most of the highway between Forks and Aberdeen crosses such wave deposited sand.

small amounts of excellent crude oil. However, the production was too small to make the wells profitable, and they were all abandoned. Wreckage of old drilling equipment still rusts in the woods, and oil still seeps into some of the old well casings. The future will bring more drilling in the lower part of the Hoh River valley.

Heavy winter waves work the rocks into this shingle beach, and gentler summer waves spread a partial covering of sand across them.

A Stretch of Coast

The road turns at the crossing of the Hoh River, and follows the stream west to its mouth. From there to Queets, about 12 miles, the highway rims the sea cliff, and the surf crashes on the beaches directly below. It is beautiful, and there are numerous places to stop and follow trails down the cliffs to the beaches.

These sea cliffs are the only easily accessible exposures of the trench filling that forms the core of the Olympic Mountains. Many exposures reveal tightly folded layers of sandstone, and in places the rock consists of angular fragments stuck together — what geologists call a "breccia." That folding and breakage happened while the sinking oceanic crust scraped these rocks into the trench.

Layers of muddy sandstone and shale deposited as turbidite sediments on the edge of the floor of the Pacific Ocean, now exposed in the sea cliffs at Kalaloch. Notice that these layers are upside down – the coarse sand lies at the top of the bed, and the darker shale at the bottom.

Quinault Lake

A large valley glacier ended at the west end of Quinault Lake, and built a morainal ridge there, which now functions as a natural dam to impound the lake. Meltwater pouring over that moraine spread over the countryside to the west, and dumped its load of sediment to build an enormous deposit of sand and gravel. The highway crosses that smooth outwash fan between Queets and Quinault. Roads on both sides of Quinault Lake lead northeast into beautiful stands of relatively uncut rain forest in the national park, one of the few easily accessible places away from clearcuts and second or third growth forest.

Grays Harbor

Between Quinault Lake and Aberdeen, U.S. 101 winds through the western part of the Willapa Hills, a landscape almost without bedrock outcrops. All of the well concealed bedrock formed on the floor of the Pacific Ocean, and consists of thick deposits of dark sandstone resting on pillow basalts.

Grays Harbor was simply the broad valley of the Chehalis River, then a roaring glacial meltwater stream, when the last ice age ended. Sea level rose as the land ice melted, and flooded the valley to make an open bay. Since then, sandspits growing north and south have nearly isolated the bay from the open ocean. However, the sand bars can not completely isolate the bay because drainage from the land will maintain an outlet to the ocean.

Winter storm waves eroded this stretched road near Ocean City, on the open coast west of Aberdeen.

GLOSSARY

Andesite: A common volcanic rock intermediate in composition between basalt and rhyolite. Andesite comes in various shades of gray ranging from very pale to quite dark, and may become green or brown through weathering or reaction with water. It is the most abundant rock in the Cascades.

Anticline: An arching fold in layered rocks.

Ash: Minute rock fragments blown from a volcano by escaping steam.

Basalt: A common volcanic rock. Basalt is fine-grained and coal black when fresh, although it may become greenish or red through reaction with water or weathering. Basalt comprises all of the oceanic crust, and is also extremely common in terrestrial volcanoes. The magma forms through partial melting of peridotite in the mantle.

Bauxite: A variety of laterite that consists mostly of aluminum oxide, and contains relatively little iron oxide or kaolinite. Bauxite is aluminum ore.

Batholith: A mass of intrusive granite that outcrops over an area greater than about 60 square miles.

Biotite: A black or dark brown variety of mica common in many igneous and metamorphic rocks. Biotite is easy to recognize by its color, and by its tendency to form flat flakes.

Dike: A sheet of igneous rock that formed as magma squirted into a fracture.

Fault: A fracture that separates blocks of the earth's crust that have slipped past each other.

Feldspar: An extremely common and abundant mineral in most kinds of igneous and metamorphic rocks. There are several kinds of feldspars, all of which tend to form blocky crystals. Plagioclase feldspars, which contain calcium and sodium, are milky white or greenish. Potassium feldspars are milky white or salmon pink.

Formation: A distinctive body of rock. Geologists name formations after geographic localities where they were first studied, or where they are especially well exposed.

Gneiss: A common metamorphic rock that typically consists mostly of feldspar, quartz, and a black mineral such as biotite, mica or hornblende. Although gneiss and granite consist of the same minerals, gneiss is banded or streaky looking, whereas granite lacks grain. Gneiss may form through metamorphism of muddy sandstones, andesite, rhyolite, or granite,

Granite: A coarse-grained igneous rock that consists mostly of feldspar and quartz. Most granites also contain black biotite mica or hornblende. Like many geologists, we use the term granite to refer broadly to a variety of more narrowly defined rocks such as granodiorite, quartz diorite, and monzonite, all of which consist largely of feldspar and quartz.

Group: Several closely related formations.

Hornblende: A black mineral common in many igneous and metamorphic rocks. Large crystals are fairly easy to recognize because they are long and narrow, and have glossy surfaces.

Igneous Rocks: Any rock that crystallizes from a melt. If the melt crystallizes below the surface, it generally becomes a fairly coarse-grained igneous rock. If it erupts through a volcano, it becomes a fine-grained volcanic rock.

Kaolinite: A clay mineral commonly used in making porcelain and fire bricks.

Laterite: A red or yellow tropical soil, which consists mostly of iron oxides, aluminum oxide, and the clay mineral kaolinite mixed in various proportions.

Lava: Molten rock that has erupted onto the earth's surface.

Lithosphere: The relatively cool and rigid outer rind of the earth, about 60 miles thick. The lower part consists of peridotite, the upper part of oceanic or continental crust.

Magma: Molten rock beneath the earth's surface. If magma erupts through a volcano, it becomes lava.

Mantle: The shell of the earth between the crust and the core. It consists of a rock called peridotite, which comprises most of the earth's volume.

Metamorphic Rocks: Rocks that have changed their form through recrystallization at high temperature, and generally under high pressure. Common metamorphic rocks include marble, gneiss, schist, phyllite, and slate.

Migmatite: A mixed rock, partly igneous, partly metamorphic. Migmatite generally forms as rocks get hot enough during metamorphism to begin to melt. Many migmatites consist of lighter gray or pink granite and dark gray gneiss swirled together in a marble cake pattern.

Moraine: A ridge composed of glacial till, which marks the former edge of a glacier.

Palagonite: A yellowish alteration product that forms through reaction of basalt with steam, commonly where basalt lava flows entered lakes.

Peat: A dark brown or black sedimentary deposit composed of partially decayed plant remains. Peat typically forms in swamps or marshes, and eventually becomes coal.

Pillow: A cylindrical mass of basalt that forms where lava flows moved under water. Common in the oceanic crust, and locally on the Columbia Plateau.

Plate: One of the dozen or so large segments of the earth's lithosphere that together tile the entire surface of the planet.

Pluton: A mass of intrusive igneous rock, most commonly granite.

Quartz: A common and abundant mineral in many kinds of igneous, sedimentary, and metamorphic rocks. Quartz typically occurs in irregular transparent grains, which look almost like little pieces of light gray glass.

Rhyolite: A common volcanic rock that forms from lava that has the same composition as granite. Rhyolite is typically some pale shade of gray, yellow, or pink. It tends to erupt violently.

Schist: A common metamorphic rock that contains enough mica to give it a flaky quality. Schist commonly forms through metamorphism of muddy sandstones.

Sedimentary Rocks: Rocks formed from deposits of detrital material such as gravel, sand, mud, silt, or peat.

Serpentinite: A soft dark green rock that forms through reaction of peridotite with water.

Sill: A sheet of igneous rock sandwiched between layers of sedimentary rocks.

Syncline: A trough folded into layered rocks.

Tectonic: Large scale movements of the earth's crust or mantle.

Terrain: Geologists use this term to refer to an assemblage of rocks that appear to share a common history.

Till: Glacially deposited sediment, which consists of debris of all sizes mixed indiscriminately together.

Trench: A deep trough in the ocean floor that forms where a lithospheric plate is sinking into the earth's interior.

Trench Filling: A chaotic assemblage of severely deformed oceanic sediments and slices of oceanic crust that were scraped into a trench.

Turbidites: Muddy sandstones laid down on the ocean floor as sediment settles from clouds of muddy water. Typical turbidites consist of layers that grade from coarse sand at the base to very fine silt or clay at the top, a reflection of the rates at which the different sediment sizes settle.

Varves: Alternating thin layers of pale silt deposited in summer and darker clay deposited in winter. The number of annual pairs records the period of deposition in a glacial lake.

SUGGESTED READING

Most of the books and reports in this short list are reasonably available and either deliberately addressed to a non-technical audience or written in a manner that people who are not geologists will not find too intimidating. We also include several more formidable references simply because they treat their subject so completely that most people will need to read no further.

Allen, John E., 1979, *The Magnificent Gateway*, Timber Press, Forest Grove, Oregon.

Baker, V.R., 1973, "Paleohydrology and Sedimentology of Lake Missoula Flooding in Eastern Washington": Geological Society of America, Special Paper 144.

Baker, V.R. and Nummedal, Dag, 1978, "The Channeled Scabland, a Guide to the Geomorphology of the Columbia Basin, Washington": Planetary Geology Program, Office of Space Science, National Aeronautics and Space Administration, Washington, D.C.

Barksdale, Julian D., 1975, "Geology of the Methow Valley, Okanogan County, Washington": Washington Division of Geology and Earth Resources, Bulletin 68.

Bretz, J.H., 1969, "The Lake Missoula Floods and the Channeled Scabland of Washington, New Data and Interpretations": Journal of Geology, v. 77, p. 503-543.

Campbell, Newell P., 1975, "A Geologic Road Log over Chinook, White Pass, and Ellensburg to Yakima Highways": Washington Division of Geology and Earth Resources, Circular 54.

Crandell, Dwight R., 1969, "Surficial Geology of Mount Rainier National Park, Washington": U.S. Geological Survey, Bulletin 1288.

Crandell, Dwight R., 1969, "The Geologic Story of Mount Rainier, A Look at the Geologic Past of One of America's Most Scenic Volcanoes": U.S. Geological Survey, Bulletin 1292.

Easterbrook, Don J. and Rahm, David A., 1970, *Landforms of Washington*, Western Washington State College, Bellingham.

Gilmour, Ernest H. and Stradling, Dale, editors, 1970, *Proceedings of the Second Columbia River Basalt Symposium, Cheney:* Eastern Washington State College Press.

Halliday, W.R., 1963, "Caves of Washington": Washington Division of Mines and Geology, Information Circular No. 40.

Harris, Stephen L., 1980, *Fire and Ice: The Cascade Volcanoes*, The Mountaineers, Seattle.

Livingston, V.E., Jr., 1959, "Fossils in Washington": Washington Division of Mines and Geology, Information Circular No. 33.

Lipman, Peter W. and Mullineaux, Donal R., editors, 1983, "The 1980 Eruptions of Mount St. Helens, Washington": U.S. Geological Survey, Professional Paper 1250.

McKee, Bates, 1972, *Cascadia: The Geologic Evolution of the Pacific Northwest*, McGraw-Hill, New York.

Snavely, P.D. and Wagner, H.C., 1963, "Tertiary Geologic History of Western Oregon and Washington": Washington Division of Mines and Geology, Report of Investigation 22.

Tabor, R.W., 1975, *Guide to the Geology of Olympic National Park:* University of Washington Press, Seattle.

Tabor, Rowland W. and Crowder, D.F., 1969, "On Batholiths and Volcanoes — Intrusion and Eruption of Late Cenozoic Magmas in the Glacier Peak Area, North Cascades, Washington": U.S. Geological Survey, Professional Paper 604.

Rau, Weldon W., 1980, "Washington Coastal Geology between the Hoh and Quillayute Rivers": Washington Division of Geology and Earth Resources, Bulletin 72.

Vance, Joseph A., 1975, "Bedrock Geology of San Juan County: in Geology and Water Resources of the San Juan Islands," R.H. Russel, editor: Washington Department of Ecology and Water Supply, Bulletin 46.

Index

We encourage you to patronize your local bookstore. Most stores will order any title they do not stock. You may also order directly from Mountain Press, using the order form provided below or by calling our toll-free, 24-hour number and using your VISA, MasterCard, Discover, or American Express.

Some geology titles of interest:

____ROADSIDE GEOLOGY OF ALASKA 18.00
____ROADSIDE GEOLOGY OF ARIZONA 18.00
____ROADSIDE GEOLOGY OF COLORADO, Second Edition 20.00
____ROADSIDE GEOLOGY OF HAWAII 20.00
____ROADSIDE GEOLOGY OF IDAHO 20.00
____ROADSIDE GEOLOGY OF INDIANA 18.00
____ROADSIDE GEOLOGY OF MAINE 18.00
____ROADSIDE GEOLOGY OF MASSACHUSETTS 20.00
____ROADSIDE GEOLOGY OF MONTANA 20.00
____ROADSIDE GEOLOGY OF NEBRASKA 18.00
____ROADSIDE GEOLOGY OF NEW MEXICO 18.00
____ROADSIDE GEOLOGY OF NEW YORK 20.00
____ROADSIDE GEOLOGY OF NORTHERN and CENTRAL CALIFORNIA 20.00
____ROADSIDE GEOLOGY OF OREGON 16.00
____ROADSIDE GEOLOGY OF PENNSYLVANIA 20.00
____ROADSIDE GEOLOGY OF SOUTH DAKOTA 20.00
____ROADSIDE GEOLOGY OF TEXAS 20.00
____ROADSIDE GEOLOGY OF UTAH 20.00
____ROADSIDE GEOLOGY OF VERMONT and NEW HAMPSHIRE 14.00
____ROADSIDE GEOLOGY OF VIRGINIA 16.00
____ROADSIDE GEOLOGY OF WASHINGTON 18.00
____ROADSIDE GEOLOGY OF WISCONSIN 18.00
____ROADSIDE GEOLOGY OF WYOMING 18.00
____ROADSIDE GEOLOGY OF THE YELLOWSTONE COUNTRY 12.00
____CHASING LAVA: A Geologist's Adventures at the Hawaiian Volcano Observatory 12.00
____EVIDENCE FROM THE EARTH: Forensic Geology and Criminal Investigation 20.00
____FINDING FAULT IN SOUTHERN CALIFORNIA: An Earthquake Tourist Guide 18.00
____GEOLOGY OF THE LAKE SUPERIOR REGION 22.00
____GEOLOGY OF THE LEWIS AND CLARK TRAIL IN NORTH DAKOTA 18.00
____GEOLOGY UNDERFOOT IN CENTRAL NEVADA 16.00
____GEOLOGY UNDERFOOT IN DEATH VALLEY AND OWENS VALLEY 16.00
____GEOLOGY UNDERFOOT IN ILLINOIS 18.00
____GEOLOGY UNDERFOOT IN SOUTHERN CALIFORNIA 14.00
____GLACIAL LAKE MISSOULA AND ITS HUMONGOUS FLOODS 15.00
____NORTHWEST EXPOSURES: A Geologic Story of the Northwest 24.00

Please include $3.00 per order to cover postage and handling.

Send the books marked above. I enclose $_____

Name_____

Address _____

City/State/Zip _____

☐ Payment enclosed (check or money order in U.S. funds)

Bill my: ☐ VISA ☐ MasterCard ☐ Discover ☐ American Express

Card No. _____ Expiration Date _____

Signature _____

MOUNTAIN PRESS PUBLISHING COMPANY
P.O. Box 2399 • Missoula, MT 59806 • Order Toll-Free 1-800-234-5308
E-mail: info@mtnpress.com • Web: www.mountain-press.com